Lecture Notes in Mathematics

Edited by A. Dold and B. Eckmann

Subseries: Instituto de Matemática Pura e Aplicada, Rio de Janeiro
Adviser: C. Camacho

1093

Alexander Prestel

Lectures on Formally Real Fields

Springer-Verlag
Berlin Heidelberg New York Tokyo 1984

Author

Alexander Prestel
Fakultät für Mathematik, Universität Konstanz
Postfach 5560, 7750 Konstanz, Federal Republic of Germany

This book was originally published in 1975 by the Instituto de Matemática Pura e Aplicada, Rio de Janeiro as volume 22 of the series "Monografias de Matemática".

AMS Subject Classification (1980): 12D15, 10C04

ISBN 3-540-13885-4 Springer-Verlag Berlin Heidelberg New York Tokyo
ISBN 0-387-13885-4 Springer-Verlag New York Heidelberg Berlin Tokyo

Printing and binding: Beltz Offsetdruck, Hemsbach / Bergstr.
2146/3140-543210

To the memory of my parents

Preface

Ten years ago, in 1974, these lectures were given at the 'Instituto de Matemática Pura e Aplicada', in Rio de Janeiro. In 1975 the notes of these lectures appeared in the series 'Monografias de Matemática' as No.22 published by IMPA. From that time on the notes served as an introductory text for the theory of real fields and its connections with valuation theory and quadratic form theory.

Since the 'Lectures on formally real fields' have been used in many publications as a standard reference and since currently they still seem to be the only introductory text to this theory, I gratefully accepted the proposal of C. Camacho to republish them in the IMPA-Subseries of the 'Lecture Notes in Mathematics'. It seems wise not to make changes in this new edition apart from correcting misprints and a few minor errors: The only change made is to replace the term q-*ordering* by *semiordering*, since the latter term turned out to be almost exclusively used in the recent literature.

For the developments during the last decade in this theory, particularly in the theory of reduced quadratic forms, I would like to refer the reader to two publications of T.Y. Lam: his survey article

The Theory of Ordered Fields [in Ring Theory and Algebra III (ed. B. Mc Donald), Lecture Notes in Pure and Applied Math., Vol. 55, Dekker, New York, 1980, p. 1-152] and to his expository notes

Orderings, Valuations and Quadratic Forms [Conf. Board of the Math. Sciences, Regional Conf. Series in Math., No. 52, Providence,R.I.,1983].

Finally I would like to thank Edda Polte for preparing the typescript of this second edition.

Konstanz, 1984 A. Prestel

CONTENTS

Introduction

In mathematics, the following method of generalization has turned out to be very successful: one starts with a given well-known mathematical structure, singles out the most basic properties of that structure (axioms) and then considers the class of all structures satisfying these properties. Finally, one tries to characterize the original structure among the members of this class.

A well known example of that method is the following: considering the set $\mathbb{C}$ of complex numbers together with the usual (field-) operations, the most basic properties are the rules for dealing with these operations. All these rules are implied by the "axioms" of fields of characteristic zero. Among this class of fields, $\mathbb{C}$ cannot be characterized completely using algebraic properties only. But there is a subclass consisting of the algebraically closed fields of characteristic zero which contains $\mathbb{C}$ and shares with $\mathbb{C}$ all "algebraic properties". This fact is usually called the "Lefschetz-Principle".

Besides $\mathbb{C}$, the field $\mathbb{R}$ of real numbers is another outstanding mathematical structure. One of the main differences between $\mathbb{C}$ and $\mathbb{R}$ is the existence of a so-called "ordering" on $\mathbb{R}$. Now, the most basic properties are the field axioms together with the rules for dealing with the ordering "$\leq$". These rules are all implied by the axioms of a linear order and

(1) $x \leq y \Rightarrow x + z \leq y + z$

(2) $x \leq y,\ 0 \leq z \Rightarrow xz \leq yz$.

A field F together with a linear ordering $\leq$ satisfying (1) and (2) is called an <u>ordered</u> field $(F,\leq)$. Among the class of ordered fields, $\mathbb{R}$ cannot be characterized completely using only algebraic and order-theoretical properties, without involving set-theoretical properties (completeness!). But once more, there is a subclass, the class of maximally ordered (or real closed) fields, which shares with $\mathbb{R}$ all algebraic and order-theoretical properties. This fact is usually called the "Tarski-Principle".

All this is contained in § 1 to § 5. To state these "Transfer-Principles" precisely and to prove them, we need some notions from model theory (introduced in § 4). The theory presented is self-contained up to the proof of the main theorem of model theory, the compactness theorem.

A field F admitting at least one ordering, in general admits several. The set X_F of all orderings of F can be given a topology in such a way that it becomes a compact and totally disconnected space. The study of such spaces is contained in § 6 to § 9.

A field admitting at least one ordering is called <u>formally real</u>. These fields are characterized algebraically by the condition: -1 is not a sum of squares in F, or equivalently, no quadratic equation of the type

$$X_1^2 + \ldots + X_n^2 = 0$$

has a non-trivial solution in F. This characterization is one of the most elementary connections between the theory of formally real fields and the theory of quadratic forms. More generally, if $\leq$ is an ordering of the field F and $a_1, \ldots, a_n \in F$ are positive with respect to $\leq$, then the quadratic equation

$$(*) \qquad a_1 X_1^2 + \ldots + a_n X_n^2 = 0$$

admits no non-trivial solution in F. This is based on the important consequence

$$(2') \qquad 0 \leq x \Rightarrow 0 \leq xy^2$$

of (2). Hence, even if $a_1, \ldots, a_n > 0$ and $\leq$ is a semiordering of F, i.e. a linear ordering satisfying (1), (2') and $0 < 1$, then (*) admits no non-trivial solution in F.

This observation indicates the usefulness of the study of semiorderings in connection with quadratic forms over formally real fields. Hence, right from the beginning we deal with semiorderings. These turn out to be of great importance in proving certain "Local-Global Principles" (see § 8).

One class of fields turns out to be of special interest. It consists of all fields in which every semiordering is already an ordering. The space X_F of orderings in this case satisfies the so-called "Strong Approximation Property". Hence these fields are called SAP-fields. This class will be studied in § 9. In § 10 SAP-fields will be characterized by their Witt rings of quadratic forms.

These notes are based on a course on formally real fields taught at IMPA during the winter period from April to June 1974. The author's stay at IMPA took place under the German-Brazilian Cooperation Agreement GMD-CNPq. The typescript was prepared by Wilson Góes. I want to thank him for the fine job he did.

Rio de Janeiro 1975 A. Prestel

§ 1. ORDERINGS AND SEMIORDERINGS OF FIELDS

Orderings

Let F be a field. An ordering $\leq$ of F is a binary relation satisfying

(1.1) (i) $a \leq a$

(ii) $a \leq b,\ \ b \leq c \Rightarrow a \leq c$

(iii) $a \leq b,\ \ b \leq a \Rightarrow a = b$

(iv) $a \leq b$ or $b \leq a$

(v) $a \leq b \Rightarrow a + c \leq b + c$

(vi) $0 \leq a,\ \ 0 \leq b \Rightarrow 0 \leq ab$.

The set $P = \{a \in F \mid 0 \leq a\}$ obviously satisfies

(1.2) (1) $P + P \subset P$

(2) $P \cdot P \subset P$

(3) $P \cap -P = \{0\}$

(4) $P \cup -P = F$.

P is called the positive cone of $\leq$ (although it includes 0). A subset $P \subset F$ satisfying (1) to (4) is called a positive cone of F . It is easy to see that, for a positive cone P,

$$a \leq b :\Longleftrightarrow b - a \in P$$

defines an ordering on F such that P is its positive cone. Hence we will sometimes call P an ordering as well. Examples of ordered fields are $\mathbb{Q}$ and $\mathbb{R}$ with their usual orderings.

To get an algebraic criterion for fields which admit some ordering, we generalize the notion of positive cones to pre-positive cones of fields. But looking forward to §5 we will go one step further and also consider rings.

Let A be a commutative ring with unit 1 and I a subset of A.

Then we call a subset P of A a <u>pre-positive cone of</u> A if

(1.3) (1) $P + P \subset P$

(2) $P \cdot P \subset P$

(3) $-1 \notin P$

(4) $A^2 \subset P$.

(1.4) LEMMA <u>Let</u> P_o <u>be a pre-positive cone of</u> A. <u>Then there is an extension</u> P <u>of</u> P_o <u>satisfying in addition</u>:

$$P \cup -P = A \quad \text{and} \quad P \cap -P \quad \text{is a prime ideal.}$$

<u>Proof</u>: By Zorn's lemma the set of pre-positive cones extending P_o contains some maximal element. Let $P \supset P_o$ be such a maximal pre-positive cone. We claim

$$Px \cap (1+P) = \emptyset \quad \text{or} \quad -Px \cap (1+P) = \emptyset$$

for all $x \in A$. Suppose we have $p_1 x = 1 + q_1$ and $-p_2 x = 1 + q_2$ for some $p_1, p_2, q_1, q_2 \in P$. Multiplying both equations we obtain $-p_1 p_2 x^2 \in 1 + P$. But then $-1 \in P$ which is a contradiction to (3).

To prove $P \cup -P = A$ let $x \in A$ and assume first $Px \cap (1+P) = \emptyset$. For $P' := P - Px$ we get $P \subset P'$, $-x \in P'$, $P' + P' \subset P'$ and $P' \cdot P' \subset P'$. If $-1 \in P' = P - Px$ then $Px \cap (1+P) \neq \emptyset$. Hence P' is again a pre-positive cone. But by the maximality of P, $P' = P$. Hence $-x \in P$. If we assume now $-Px \cap (1+P) = \emptyset$ we get $x \in P$ by the same argument. It remains to show that $J := P \cap -P$ is a prime ideal of A. From $P \cup -P = A$ it follows easily that J is an ideal. Now let $a_1 \cdot a_2 \in J$. We may assume without loss of generality $-a_1, -a_2 \notin P$. But then from the above argument we get $Pa_1 \cap (1+P) \neq \emptyset$ and $Pa_2 \cap (1+P) \neq \emptyset$. Let $p_1 a_1 = 1 + q_1$ and $p_2 a_2 = 1 + q_2$ for some $p_1, p_2, q_1, q_2 \in P$. Multiplying we obtain $p_1 p_2 a_1 a_2 \in 1 + P$. Hence $-1 \in P - p_1 p_2 a_1 a_2 \subset P + J \subset P$ gives a contradiction.

q.e.d.

In case A is a field F, we call $P \subset F$ a <u>pre-positive cone</u> of F if it satisfies (1.3) (1) to (4). Obviously any positive cone P of F is also a pre-positive cone, since for any $x \in F$ we have $x \in P$ and hence $x^2 \in P$ or $-x \in P$ and hence $x^2 = (-x)(-x) \in P$. In particular we have $1^2 = 1 \in P$, which implies $-1 \notin P$.

(1.5) COROLLARY <u>Any pre-positive cone P_o of a field F can be extended to some positive cone P of F.</u>

From the proof of Lemma (1.4) we get another

(1.6) COROLLARY <u>Any pre-positive cone P_o of a field F equals the intersection of all positive cones P extending P_o.</u>

<u>Proof</u>: Obviously P_o is included in the intersection. Now let $x \notin P_o$. But then $P_o x \cap (1+P_o) = \emptyset$, since from $px = 1 + q$ and $p,q \in P_o$ we would get $x = (1+q)\, p(\frac{1}{p})^2 \in P_o$. As in the proof of (1.4), $P' = P_o - x P_o$ is a pre-positive cone extending P_o and containing $-x$. By Corollary (1.5), P' can be extended to a positive cone of F. Hence x is not in the intersection.

q.e.d.

Now let

$$S_F = \text{set of sums of squares of } F .$$

(1.7) PROPOSITION (a) <u>S_F is contained in every pre-positive cone of</u> F.

(b) <u>S_F is closed under addition</u>.

(c) <u>$\dot{S}_F$ is a multiplicative subgroup of</u> $\dot{F}$.

<u>Proof</u>: (a) and (b) are trivial.

(c) $\Sigma\, a_i^2 \neq 0 \Rightarrow (\Sigma\, a_i^2)^{-1} = \Sigma\, \left(\frac{a_i}{\Sigma\, a_i^2}\right)^2$.

q.e.d.

The following theorem gives an algebraic characterization of fields which admit some ordering. These fields will be called <u>formally real</u> or <u>orderable</u>.

(1.8) THEOREM <u>For a field</u> F , (a) <u>to</u> (d) <u>are equivalent</u>:

(a) F <u>is formally real</u>

(b) $-1 \notin S_F$

(c) $\Sigma a_i^2 = 0 \Rightarrow$ <u>all</u> $a_i = 0$

(d) $F \neq S_F$.

<u>Proof</u>: (b) $\Leftrightarrow$ (c) and (a) $\Rightarrow$ (d) are easy to see.

(d) $\Rightarrow$ (b): If $-1 \in S_F$, by $a = (\frac{a+1}{2})^2 + (-1)(\frac{a-1}{2})^2$, $a \in S_F$ for any $a \in F$, i.e. $S_F = F$.

(b) $\Rightarrow$ (a): If $-1 \notin S_F$, S_F is a pre-positive cone of F , hence by Corollary (1.5) can be extended to a positive cone of F .

q.e.d.

(1.9) COROLLARY $S_F = \bigcap_{P \text{ positive cone}} P$.

<u>Proof</u>: If F is not orderable, $S_F = F$ and the intersection over the empty index set is by definition F . If F is orderable, this equation is a special case of Corollary (1.6).

q.e.d.

Hence the elements of S_F are also called <u>totally positive</u>.

<u>Remark</u>: The characteristic of a formally real field F is obviously zero. Hence the field $\mathbb{Q}$ of rational numbers is isomorphic to a subfield of F .

(1.10) PROPOSITION <u>Let</u> P_1, P_2 <u>be positive cones of a field</u> F . <u>Then</u> $P_1 \subset P_2$ <u>implies</u> $P_1 = P_2$.

Proof: Let $x \in P_2$ and $x \notin P_1$. But then $-x \in P_1 \subset P_2$ gives a contradiction.

q.e.d.

(1.11) LEMMA S_F is a positive cone of a field F iff F has a unique ordering.

Proof: "$\Rightarrow$" by Proposition (1.10).

"$\Leftarrow$" suppose $S_F \cup -S_F \neq F$. Let $x, -x \notin S_F$. Then by Corollary (1.9) there are positive cones P_1, P_2

$x \notin P_1$, $-x \notin P_2$. But then $-x \in P_1$, $x \in P_2$ gives a contradiction.

q.e.d.

Semiorderings

As we remarked in the introduction there is a certain generalization of positive cones and orderings which is based on the following observation. Very often one only uses the property

$$0 \leq a \Rightarrow 0 \leq ab^2$$

of an ordering together with $0 < 1$. This is especially the case if one deals with quadratic forms. Hence we will call a binary relation $\leq$ of a field F a semiordering if it satisfies (1.1) (i) to (v) and

(vi!) $0 < 1$

$0 \leq a \Rightarrow 0 \leq ab^2$

for all $a,b \in F$. The set $P = \{a \in P \mid 0 \leq a\}$ then satisfies

(1) $P + P \subset P$

(2) $F^2 \cdot P \subset P$ and $1 \in P$

(3) $P \cap -P = 0$

(4) $P \cup -P = F$.

Such a set will be called a <u>semicone</u> of F . Obviously a semicone $P \subset F$ determines a semiordering $\leq$ of F defined by

$$a \leq b :\Longleftrightarrow b - a \in P .$$

Hence we will sometimes call P a semiordering as well.

Any ordering of F is also a semiordering. But there are semi-orderings $\leq$ of some fields F which are not orderings (as will be seen in §6), i.e. there are $a,b \in F$ such that $0 < a,b$ and $ab < 0$. A semiordering which is not an ordering will be called a <u>proper semi-ordering</u>.

For semiorderings the statements corresponding to (1.5) to (1.11) also hold. To see this let us first introduce the notion of a pre-semicone, which is not quite analogous to that of a pre-positive cone. We call a subset $P \subset F$ a <u>pre-semicone</u> if it satisfies

(1) $P + P \subset P$

(2) $F^2 \cdot P \subset P$

(3) $P \cap -P = \{0\}$.

Nothing is said about 1. Obviously any semicone is a pre-semicone. (2) may be replaced by $S_F \cdot P \subset P$.

(1.12) LEMMA If P is a <u>pre-semicone of a formally real field</u> F <u>and</u> $x \notin P$ <u>then there is a pre-semicone</u> P' <u>extending</u> P <u>and containing</u> $-x$.

<u>Proof</u>: Let $P' = P - x\,S_F$. Obviously P' satisfies (1) and (2). Let $p - xs \in -P'$ for some $p \in P$, $s \in S_F$. Then there are $p_1 \in P$, $s_1 \in S_F$ such that $(p+p_1) - x(s+s_1) = 0$. If $s + s_1 \neq 0$, $x = (p+p_1)(s+s_1)^{-1} \in P$ gives a contradiction. Hence $s + s_1 = 0$. Since F is formally real, $s = s_1 = 0$ by (1.8) (c). Hence $p + p_1 = 0$ But then $p \in P \cap -P$, which implies $p = p_1 = 0$.

q.e.d.

(1.13) LEMMA *Any pre-semicone* P_o *of a formally real*[1] *field* F *can be extended to some* P *such that* P *or* $-P$ *is a semicone of* F.

Proof: By Zorn's lemma the set of pre-semicones extending P_o contains some maximal pre-semicone P. If then $x \notin P$, by Lemma (1.12) there is an extension P' of P such that $-x \in P'$. From the maximality of P we get $-x \in P$. Hence $P \cup -P = F$. Now P or $-P$ is a semicone.

q.e.d.

(1.14) COROLLARY *If* F *is formally real*, S_F *equals the intersection of all semicones*.

Proof: Follows directly from (1.12) and (1.13). Note that S_F is a pre-semicone containing 1 if F is formally real.

q.e.d.

(1.15) COROLLARY F *is formally real iff it admits a semiordering*.

By the same argument as in (1.10) we get:

(1.16) PROPOSITION *If* $P_1 \subset P_2$ *are both semicones of* F, *then* $P_1 = P_2$.

(1.17) COROLLARY *If* F *has a unique (semi)ordering, it has no proper semiordering*.

Proof: By the same argument as in (1.11) the assumption implies that S_F forms a positive cone of F. Now any semicone P contains S_F. Hence $P = S_F$ by (1.16).

q.e.d.

---

[1] Actually "formally real" is superfluous. For a pre- semicone P_o, $S_F \subset \pm P_o$ and hence $-1 \notin S_F$.

We will now prove some properties of semiorderings and look at the semiorderings of algebraic number fields, i.e. finite algebraic extensions of $\mathbb{Q}$.

(1.18) LEMMA *Let* $\leq$ *be a semiordering of a field* F . *Then for every* $a,b \in F$

(1) $0 < a \Rightarrow 0 < \frac{1}{a}$

(2) $0 < a < b \Rightarrow ba^2 < ab^2$

(3) $0 < a < b \Rightarrow \frac{1}{b} < \frac{1}{a}$

(4) $1 < b \Rightarrow b < b^2$

(5) $0 < a < 1 \Rightarrow a^2 < a$

(6) $0 < a < b\ ,\ a \in S_F \Rightarrow a^2 < b^2$

(7) $0 < a < b\ ,\ b \in S_F \Rightarrow a^2 < b^2$.

Proof:

(1): $0 < a \Rightarrow 0 < a(\frac{1}{a})^2 = \frac{1}{a}$.

(2): $0 < a < b \Rightarrow 0 < a\ ,\ 0 < b-a$

$$\Rightarrow 0 < \frac{1}{\frac{1}{b-a}+\frac{1}{a}} \cdot b^2 = ab^2 - ba^2$$

$$\Rightarrow ba^2 < ab^2 \ .$$

(3): multiply (2) by $(\frac{1}{ab})^2$.

(4): put $a = 1$ in (2) .

(5): put $b = 1$ in (2) .

(6): $0 < a < b \Rightarrow ba^2 < ab^2$ by (2)

$\Rightarrow ba < b^2$ since $a^{-1} \in S_F$,

also $a < b \Rightarrow a^2 < ba$ since $a \in S_F$.

Finally we get $a^2 < b^2$.

(7): similar to (6). q.e.d.

Remark: A semiordering $\leq$ is an ordering of F iff for all $a,b \in F$, $0 < a < b \Rightarrow a^2 < b^2$.

Archimedean Orderings and Semiorderings

We call a semiordering $\leq$ <u>archimedean</u> iff for every $a \in F$ there is an $n \in \mathbb{N}$ such that $a < n$.

(1.19) PROPOSITION <u>A semiordering $\leq$ is archimedean iff $\mathbb{Q}$ is dense in F with respect to $\leq$ (i.e. if $a < b$ then there is $r \in \mathbb{Q}$ such that $a < r < b$).</u>

<u>Proof</u>: "⇐" choose $b = a+1$.

"⇒" $a < b$ implies $0 < (b-a)^{-1}$. Let $n \in \mathbb{N}$ such that $(b-a)^{-1} < n$. By (1.18)(3) we get $0 < \frac{1}{n} < (b-a)$. Now let $m \in N$ be the first natural number $> na$. Then $m < nb$, since otherwise $n(b-a) = nb - na \leq 1$. But this implies $b-a \leq \frac{1}{n}$, since $n^{-1} \in S_F$. Now from $na < m < nb$ we get $a < \frac{m}{n} < b$, by $n^{-1} \in S_F$.

q.e.d.

(1.20) THEOREM <u>Every archimedean semiordering is an ordering</u>.

<u>Proof</u>: Let $0 < a$ and $0 < b$ and without loss of generality $a < b$. Since $b-a < b+a$, there is an $r \in \mathbb{Q}$ such that we have

$$0 < (b-a) < r < (b+a).$$

By (1.18)(6) and (7) we obtain

$$(b-a)^2 < r^2 < (b+a)^2.$$

This implies $0 < 4 \cdot ab$, or $0 < ab$ using $4^{-1} \in S_F$.

q.e.d.

(1.21) THEOREM <u>Every semiordering of an algebraic number field F is an ordering</u>.

<u>Proof</u>: Let $\leq$ be a semiordering of F. We show that $\leq$ is archimedean. Then by (1.20) it is an ordering. Let $a \in F$ and without loss of generality $1 < a$. Then by (1.18)(4) we obtain

$$1 < a < a^2 .$$

Multiplying by a^2 we get

$$a^2 < a^3 < a^4 .$$

Continuing multiplication by a^2 we obtain

$$1 < a < \ldots < a^m < a^{m+1} < \ldots$$

By (1.18)(3), this implies $0 < \frac{1}{a^m} \leq 1$ for all $m \in \mathbb{N}$. Now a is a zero of some polynomial, i.e.

$$a^n + r_{n-1} a^{n-1} + \ldots + r_o = 0$$

for some $r_{n-1}, \ldots, r_o \in \mathbb{Q}$. This implies

$$-a = r_{n-1} + \ldots + r_o \frac{1}{a^{n-1}} .$$

Hence

$$a \leq \sum_{m \leq n-1} |r_{n-1-m} \frac{1}{a^m}| \leq \sum_{m \leq n-1} |r_m| .$$

This shows that there is some $k \in \mathbb{N}$ such that $a < k$.

q.e.d.

<u>Remark</u>: For a semiordering, $|ab| = |a||b|$ in general does not hold, but $|ab| = ||a||b||$. If $a \in S_F$ then obviously $|ab| = a|b|$.

Now let us look for a moment at the most important ordered field, the field $\mathbb{R}$ of real numbers. We call a field F <u>cut-complete</u> with respect to an ordering $\leq$, if for all non-empty subsets $A, B \subset F$ such that $a \leq b$ for all $a \in A$ and $b \in B$ there is a $c \in F$ such that $a \leq c \leq b$ for all $a \in A$ and $b \in B$. This is obviously equivalent to: any non empty bounded subset of F has a least upper bound. As is well known $\mathbb{R}$ is cut-complete. We will show that $\mathbb{R}$ is up to isomorphism the only cut-complete field and that any archimedean ordered field can be embedded into the reals.[2)]

(1.22) PROPOSITION *If* F *is cut-complete with respect to the ordering* $\leq$ *, then* $\leq$ *is archimedean.*

Proof: Assume the set B of upper bounds of $\mathbb{N}$ to be not empty. Since $n < b$ for all $n \in \mathbb{N}$ and $b \in B$, there is $c \in F$ such that $n \leq c \leq b$ for all $n \in \mathbb{N}$ and $b \in B$. Hence $c \in B$. But also $c-1 \in B$, contradicting $c \leq b$ for all $b \in B$.

q.e.d.

(1.23) THEOREM *Let* F *be cut-complete with respect to an ordering* $\leq$.

(a) *Any archimedean ordered field admits an order-preserving isomorphism to* F.

(b) F *is order-isomorphic to* $\mathbb{R}$.

Proof: Let F_1 be archimedean ordered by $\leq_1$. $\mathbb{Q}$ is isomorphic to a dense subfield Q of F and to a dense subfield Q_1 of F_1. Hence we have an order-preserving isomorphism $l : Q_1 \to Q$. This isomorphism may be extended as follows. Any $c_1 \in F_1$ determines the subsets $A_1 = \{a \in Q_1 \mid a \leq_1 c_1\}$ and $B_1 = \{b \in Q_1 \mid c_1 \leq_1 b\}$. Since F is cut-complete there is some $c \in F$ such that $l(a) \leq c \leq l(b)$ for all $a \in A_1$ and $b \in B_1$. The density of Q in F implies the uniqueness of c. Hence we may define $l(c_1) := c$. One easily checks that l is an order-isomorphism of F_1 into F. This proves (a).

In case F_1 itself is cut-complete (e.g. $F_1 = \mathbb{R}$) then l is easily seen to be onto. This proves (b).

q.e.d.

---

2) One can also define "cut-complete" for semiordered fields. But as will be seen in the following proposition this semiordering is archimedean and hence an ordering.

Remark: Since every field which admits an archimedean ordering can be embedded into $\mathbb{R}$ one could call these fields "real". But usually the word "real" stands for "formally real", as will be the case in the following.

Extensions of Orderings and Semiorderings

Next let us discuss extensions of (semi)orderings of a field F to larger fields.

(1.24) LEMMA *Let* $F \subset F_1$ *be fields. Then a semicone (positive cone)* P *of* F *extends to some semicone (positive cone)* P_1 *of* F_1 *iff* $\sum_{i=1}^{n} a_i X_i^2$ *has no non trivial solution in* F_1 *for all* $n \in \mathbb{N}$, $a_1,\dots,a_n \in \dot{P}$.

Proof: "$\Rightarrow$" trivial.

"$\Leftarrow$" Let us first consider the case of a semicone P of F. Consider the set

$$P_0 = \{\sum_{i=1}^{n} a_i v_i^2 \mid n \in \mathbb{N},\ a_1,\dots,a_n \in \dot{P},\ v_1,\dots,v_n \in F_1\}.$$

Obviously $P_0 + P_0 \subset P_0$ and $F_1^2 \cdot P_0 \subset P_0$. Suppose $\sum a_i v_i^2 \in -P_0$ for some $a_i \in \dot{P}$. Then there are $b_j \in \dot{P}$ and $w_j \in F_1$ such that $\sum a_i v_i^2 + \sum b_j w_j^2 = 0$. From the assumption we get $a_i = b_j = 0$. Hence $P_0 \cap -P_0 = \{0\}$, which shows that P_0 is a pre-semicone of F_1. Now by Lemma (1.13) P_0 extends to a semicone of F_1, since $1 \in P \subset P_0$.

In case P is an ordering of F, $P_0 \cdot P_0 \subset P_0$ also holds. Thus P_0 is a pre-positive cone of F_1 (since $P_0 \cap -P_0 = \{0\}$ and $1 \in P_0$, $-1 \notin P_0$). Hence by Corollary (1.5), P_0 can be extended to a positive cone of F_1.

q.e.d.

(1.25) LEMMA (Springer [Sp]) Let F_1 be an algebraic extension field of F of odd degree and let $a_1,\dots,a_n \in \dot{F}$. If $\sum_{i=1}^{n} a_i X_i^2$ has a non-trivial solution in F_1, it also has one in F.

Proof: Without loss of generality let $F_1 = F(\alpha)$ and $g \in F[X]$ be the minimal polynomial of α over F. Let $\deg g = 2n+1$ and $n \neq 0$. Since $\Sigma a_i X_i^2$ has a non-trivial solution in F_1, there are $f_1,\dots,f_n \in F[x]$, not all zero, such that $\deg f_i \leq 2n$ and

$$\Sigma a_i f_i^2(x) = g(x)\, h(x)$$

for some $h \in F[x]$. Suppose now that $\Sigma a_i X_i^2$ has no non-trivial solution in F. Then it follows that $\deg \Sigma a_i f_i^2$ is even. Hence $\deg h$ is odd and $\leq 2\cdot(2n) - (2n+1) = 2n-1$. Thus there is an irreducible factor h_1 of h of odd degree $\leq 2n-1$. Now consider $F_2 := F(\beta)$ where β is a zero of h_1. But then $\Sigma a_i X_i^2$ has a non-trivial solution in F_2, since we may assume $f_1,\dots,f_n$ to have no common divisor. Since $[F_2:F] = \text{odd}$ and $< [F_1:F]$ we can repeat this procedure till we reach a field F' such that $[F':F] = 1$, i.e. $F' = F$. But this contradicts our assumption.

q.e.d.

(1.26) THEOREM Let P be a semicone (positive cone) of F. Then P can be extended to a semicone (positive cone) of F_1 in the following cases:

(a) $F_1 = F(\sqrt{a})$ and $a \cdot P \subset P$ for some $a \in F$

(b) $[F_1:F] = \text{odd}$.

Proof: We use the criterion of Lemma (1.24). Hence let $a_1,\dots,a_n \in \dot{P}$ and let $\Sigma a_i X_i^2$ have a non-trivial solution in F_1.

(a) If $F_1 = F(\sqrt{a})$ we get

$$0 = \Sigma a_i (b_i + c_i \sqrt{a})^2$$

for some $b_i, c_i \in F$, not all zero. In particular we obtain

$$0 = \Sigma\, a_i\, b_i^2 + \Sigma\, a\, a_i\, c_i^2 .$$

Since $a_i \in \dot{P}$ and $a \cdot P \subset P$, also $a\, a_i \in \dot{P}$. But then $b_i = c_i = 0$, which gives a contradiction.

(b) If $[F_1 : F] = $ odd, $\Sigma\, a_i\, X_i^2$ has a non-trivial solution in F by Lemma (1.25), contradicting $a_i \in \dot{P}$.

q.e.d.

<u>Remark</u>: In case P is an ordering, $a \cdot P \subset P$ is equivalent to $a \in P$.

In §3 we shall call an ordered field maximally ordered if it admits no algebraic order-extension field. In §3 we will prove one of the most fundamental theorems in the theory of ordered fields, a theorem of Artin and Schreier, stating that any ordered field has an (up to isomorphism) unique algebraic order-extension field, which is maximally ordered. To prove this theorem we will first introduce some basic notions and theorems on quadratic forms.

§ 2. QUADRATIC FORMS OVER FORMALLY REAL FIELDS

(Part I)

Quadratic Forms

First we will give a short introduction to the basic notions of the theory of quadratic forms over fields (of characteristic different from 2). The reader interested in the algebraic theory of quadratic forms over fields is referred to Lam's book [La].

A quadratic form over a field F is a homogeneous polynomial of degree 2. In general it has the form

$$\rho(X_1,\ldots,X_n) = \sum_{1\le i,j\le n} b_{ij} X_i X_j$$

n is called the dimension of this quadratic form. To make its coefficients symmetric we may rewrite it as

$$\rho(X_1,\ldots,X_n) = \sum_{1\le i,j\le n} \frac{1}{2}(b_{ij}+b_{ji})X_i X_j .$$

Hence the quadratic form $\rho(X_1,\ldots,X_n)$ determines uniquely a symmetric (n,n)-matrix (a_{ij}) with $a_{ij} = \frac{1}{2}(b_{ij}+b_{ji})$. Conversely, a symmetric (n,n)-matrix (a_{ij}) determines uniquely the quadratic form $\sum_{1\le i,j\le n} a_{ij} X_i X_j$. Therefore we will use the expression "quadratic form" for (a_{ij}) as well. If we treat X as the column vector $\begin{pmatrix} X_1 \\ \vdots \\ X_n \end{pmatrix}$, then we get

$$\rho(X_1,\ldots,X_n) = (X_1,\ldots,X_n)(a_{ij})\begin{pmatrix} X_1 \\ \vdots \\ X_n \end{pmatrix} = X^T(a_{ij})X,$$

where T means "transpose".

In the theory of quadratic forms, one of the most important problems is:

"when is there a non-trivial solution of

$$\rho(X_1,\ldots,X_n) = \sum_{1\le i,j\le n} a_{ij}\, X_i\, X_j = 0$$

in F." If there is one, ρ is called <u>isotropic</u>. To answer this question one first reduces $\rho(X_1,\ldots,X_n)$ by a regular linear substitution for $X_1,\ldots,X_n$ to a certain normal form. If M is an invertible (m,n)-matrix, we may make the substitution

$$\begin{pmatrix} X_1 \\ \vdots \\ X_n \end{pmatrix} = M \begin{pmatrix} Y_1 \\ \vdots \\ Y_n \end{pmatrix} .$$

Then we get the quadratic form $Y^T(b_{ij})Y = Y^T M^T(a_{ij})MY$ in the variables $Y_1,\ldots,Y_n$. Hence we define two quadratic forms (a_{ij}) and (b_{ij}) of dimension n to be <u>equivalent</u> if there is an invertible (n,n)-matrix M such that

$$(b_{ij}) = M^T(a_{ij})\ M .$$

In this case we write $(b_{ij}) \simeq (a_{ij})$. Obviously $\simeq$ is an equivalence relation.

(2.1) EXAMPLE Let us consider the 2-dimensional quadratic form

$$X_1X_2 = \tfrac{1}{2}\ X_1X_2 + \tfrac{1}{2}\ X_2X_1 = (X_1,X_2) \begin{pmatrix} 0 & \frac{1}{2} \\ \frac{1}{2} & 0 \end{pmatrix} \begin{matrix} X_1 \\ X_2 \end{matrix}$$

By the substitution $\begin{pmatrix} Y_1 \\ Y_2 \end{pmatrix} = \begin{pmatrix} 1 & 1 \\ 1 & -1 \end{pmatrix}\begin{pmatrix} X_1 \\ X_2 \end{pmatrix}$, we obtain

$$(Y_1,Y_2)\begin{pmatrix} 1 & 1 \\ 1 & -1 \end{pmatrix}\begin{pmatrix} 0 & \frac{1}{2} \\ \frac{1}{2} & 0 \end{pmatrix}\begin{pmatrix} 1 & 1 \\ 1 & -1 \end{pmatrix}\begin{pmatrix} Y_1 \\ Y_2 \end{pmatrix} = (Y_1,Y_2)\begin{pmatrix} 1 & 0 \\ 0 & -1 \end{pmatrix}\begin{pmatrix} Y_1 \\ Y_2 \end{pmatrix} = Y_1^2 - Y_2^2 .$$

Now let V be an n-dimensional vector space over F with base vectors $e_1,\ldots,e_n$. Then the n-dimensional quadratic form

$$\rho(X_1,\ldots,X_n) = \Sigma\, a_{ij}\, X_i\, X_j\ ,\quad a_{ij} = a_{ji}$$

induces a symmetric bilinear form $b_\rho^{e_1\ldots e_n}$ on V by defining

$$b_\rho^{e_1\ldots e_n}(\bar{v},\bar{w}) = \Sigma\, a_{ij}\, v_i\, w_j$$

where $\bar{v} = \Sigma\, v_i\, e_i$ and $\bar{w} = \Sigma\, w_i\, e_i$. A vector space V together with a symmetric bilinear form b is often called an <u>inner product space</u>. Two inner product spaces $(V_1,b_1),(V_2,b_2)$ are called <u>isometric</u> if there is a vector space isomorphism τ from V_1 to V_2 respecting the bilinear form, i.e. $b_2(\tau(\bar{v}),\ \tau(\bar{w})) = b_1(\bar{v},\bar{w})$ for all $\bar{v},\bar{w} \in V_1$.

Each base $e_1,\ldots,e_n$ of an inner product space (V,b) of dimension n determines uniquely the quadratic form $(b(e_i,e_j))$.

(2.2) LEMMA <u>There is a one to one correspondence between equivalence classes of n-dimensional quadratic forms (or symmetric (n,n)-matrices) and the isometry classes of n-dimensional inner product spaces</u>.

Now let (V,b) be an inner product space of dimension n . Two vectors $\bar{v},\bar{w} \in V$ are called <u>orthogonal</u> if $b(\bar{v},\bar{w}) = 0$. For any subspace W of V ,

$$W^\perp = \{\bar{v} \mid b(\bar{v},\bar{w}) = 0 \text{ for all } \bar{w} \in W\}$$

is called the <u>orthogonal complement</u> of W . The subspace $V^\perp$ of V is called the <u>radical</u> of (V,b). The inner product b is called <u>regular</u> if $V^\perp = \{0\}$, i.e. no non-zero vector is orthogonal to all vectors. A vector $\bar{v}$ is called <u>isotropic</u> if $\bar{v} \neq \bar{0}$ and $b(\bar{v},\bar{v}) = 0$.

(2.3) THEOREM (Diagonalzation) <u>Every inner product space admits an orthogonal base, i.e. a base of pairwise orthogonal vectors</u>.

<u>Proof</u>: Let W be a complementary subspace of $V^\perp$ in V , i.e. $V = W \oplus V^\perp$. Choose any base of $V^\perp$. Now consider $b_1 := b \mid W$. Obviously b_1 is a regular inner product of W . Now we proceed by induction on the dimension of W . If $\dim W = 0$, we are done.

Hence we assume $\dim W = n+1$. Let $\bar{v} \in W$ be non-isotropic. There is at least one non-isotropic vector in W. Otherwise,

$$b_1(\bar{v},\bar{w}) = \frac{1}{2}[b_1(\bar{v}+\bar{w},\bar{v}+\bar{w}) - b_1(\bar{v},\bar{v}) - b_1(\bar{w},\bar{w})] = 0$$

would contradict the regularity of b_1. Now let $U := (F\bar{v})^{\perp}$. From $b(\bar{v},\bar{v}) \neq 0$ and

$$b(\bar{v},\bar{w} - \frac{b(\bar{v},\bar{w})}{b(\bar{v},\bar{v})}\bar{v}) = 0$$

we get $U \oplus F\bar{v} = W$. Hence $\dim U = n$, and obviously, $b_1|U$ is again regular. But then by induction, $\bar{v}$ completes an orthogonal base of U to one of W.

q.e.d.

The proof of the diagonalization theorem really gives more, namely

(2.4) COROLLARY <u>Let</u> (V,b) <u>be an inner product space and</u> $\bar{v} \in V$ <u>non-isotropic. Then there is an orthogonal base of</u> V <u>including</u> $\bar{v}$.

Now let us pass to a quadratic form (symmetric matrix $\rho = (a_{ij})$). We say "ρ <u>represents</u> $c \in F$" if $\Sigma\, a_{ij} v_i v_j = c$ for some $v_i \in F$. Since the matrix $(b(e_i,e_j))$ with respect to an orthogonal base $e_1,\dots,e_n$ is of diagonal form, we get from Corollary (2.4)

(2.5) COROLLARY <u>Every</u> n-<u>dimensional quadratic form</u> (a_{ij}) <u>representing some</u> $c \in \dot{F}$ <u>is equivalent to some diagonal form</u>

$$\begin{pmatrix} c & & & \\ & a_2 & & \\ & & \ddots & \\ & & & a_n \end{pmatrix}, \quad a_i \in F .$$

In the following, we abbreviate $\begin{pmatrix} a_1 & & 0 \\ & \ddots & \\ 0 & & a_n \end{pmatrix}$ by $\langle a_1,\dots,a_n\rangle$. The notion of regularity of an inner product space is invariant under isometries. Hence we call an n-dimensional quadratic form $\rho = (a_{ij})$ regular if some corresponding inner product space $(V, b_\rho^{e_1 \ldots e_n})$ is regular. Obviously an inner product space (V,b) is regular iff it has an orthogonal base of non-isotropic vectors. Hence a quadratic form (a_{ij}) is regular iff it is equivalent to some $\langle a_1,\ldots,a_n\rangle$ such that all $a_i \neq 0$, i.e. iff $\det\langle a_1,\ldots,a_n\rangle \neq 0$. But this is equivalent to $\det(a_{ij}) \neq 0$. Note that $\det M \neq 0$ implies

$$\det M^T(a_{ij})M \equiv \det(a_{ij}) \bmod \dot{F}^2 .$$

We define $d(\rho) := \det \rho \cdot \dot{F}^2$ to be the determinant of the quadratic form $\rho = (a_{ij})$. It is invariant under equivalence of quadratic forms.

Now let $\rho_1 = \langle a_1,\ldots,a_n\rangle$ and $\rho_2 = \langle b_1,\ldots,b_m\rangle$. Then the sum and product of ρ_1 and ρ_2 are defined by

$$\rho_1 \perp \rho_2 := \langle a_1,\ldots,a_n, b_1,\ldots,b_m\rangle$$
$$\rho_1 \otimes \rho_2 := \langle a_1b_1,\ldots,a_nb_1, a_1b_2,\ldots,a_nb_m\rangle .$$

It is easy to show that these definitions are compatible with the equivalence relation $\simeq$, i.e. if ρ_1' and ρ_2' are also of diagonal form, then $\rho_1 \simeq \rho_1'$ and $\rho_2 \simeq \rho_2'$ imply $\rho_1 \perp \rho_2 \simeq \rho_1' \perp \rho_2'$ and $\rho_1 \otimes \rho_2 \simeq \rho_1' \otimes \rho_2'$. Hence we can define sum and product of equivalence classes of quadratic forms via their diagonalizations.

Remark: The equivalence classes of quadratic forms together with the operations $\perp$ and $\otimes$ form a semiring (i.e. it is a semigroup with respect to $\perp$ and also $\otimes$, and distributivity holds). The zero element is the zero-dimensional (empty) form, the unit element is $\langle 1\rangle$.

For any natural number n let

$$n\rho := \rho \perp \ldots \perp \rho \quad \text{(n-times)}.$$

Connections with Semiorderings

We will now consider quadratic forms over a formally real field F.

Let P be a semiordering of F . We then define the <u>signature</u> with respect to P of an n-dimensional quadratic form ρ by

$$\mathrm{sgn}_P\, \rho := \text{ number of } a_i \in \dot{P} - \text{ number of } a_i \in -\dot{P} \,,$$

if $\rho \simeq \langle a_1,\dots,a_n\rangle$ for some $a_i \in F$. This definition is independent of the choice of the diagonalization, as is shown in the the following theorem.

(2.6) THEOREM [3] *If* $\langle a_1,\dots,a_n\rangle \simeq \langle b_1,\dots,b_n\rangle$ *then* $\mathrm{sgn}_P\langle a_1,\dots,a_n\rangle = \mathrm{sgn}_P\langle b_1,\dots,b_n\rangle$ *for all semiorderings* P *of the field* F .

<u>Proof</u>: Let V be an n-dimensional vector space (e.g. $V = F^n$) over F. Then there are orthogonal bases $e_1,\dots,e_n$ and $e_1',\dots,e_n'$ of V such that

$$b := b^{e_1 \,\dots\, e_n}_{\langle a_1,\dots,a_n\rangle} = b^{e_1' \,\dots\, e_n'}_{\langle b_1,\dots,b_n\rangle} \,.$$

Hence $b(e_i,e_i) = a_i$, $b(e_i',e_i') = b_i$, and $b(e_i,e_j) = b(e_i',e_j') = 0$ for all $i \neq j$. Without loss of generality we may assume the considered quadratic form to be regular. We may also assume $a_i > 0$ for all $1 \leq i \leq r$ and $b_i > 0$ for all $1 \leq i \leq s$. We claim $e_1,\dots,e_r, e_{s+1}',\dots,e_n'$ to be linearly independent. Suppose they are linearly dependent, i.e. there are $\alpha_1,\dots,\alpha_r, \alpha_{s+1}',\dots,\alpha_n' \in F$, not all zero, such that $\alpha_1 e_1 +\dots+ \alpha_r e_r = \alpha_{s+1}' e_{s+1}' +\dots+ \alpha_n' e_n'$. Taking the inner product of each side with itself one gets

$$\alpha_1^2 a_1 +\dots+ \alpha_r^2 a_r = \alpha_{s+1}'^2 b_{s+1} +\dots+ \alpha_n'^2 b_n \,.$$

---

[3] In case $F = \mathbb{R}$ this is Sylvester's Law of Inertia.

This is a contradiction since the left hand side is positive while the right one is negative. From this linear independence we get

$$r + (n-s) \leq n = \dim V .$$

Hence $r \leq s$. Similarly we also get $s \leq r$.

q.e.d.

If P is a semiordering of a field F and $a_1,\ldots,a_n \in \dot{P}$, then obviously the (regular) quadratic form $\rho = \langle a_1,\ldots,a_n \rangle$ is anisotropic. Moreover, $m\rho$ is not isotropic for all $m \in \mathbb{N}$. A quadratic form ρ such that $m\rho$ is anisotropic for all $m \in \mathbb{N}$ will be called strongly anisotropic. On the contrary, ρ will be called weakly isotropic if there is some $m \in \mathbb{N}$ such that $m\rho$ is isotropic.

(2.7) PROPOSITION Let F be a formally real field. A quadratic form $\rho = \langle a_1,\ldots,a_n \rangle$ is weakly isotropic iff there are $s_1,\ldots,s_n \in S_F$, not all zero, such that $\Sigma\, a_i s_i = 0$.

Proof: If $m\rho$ is isotropic for some $m \geq 1$, then there are $v_{ij} \in F$, not all zero, such that

$$0 = \sum_{j=1}^{m} \left(\sum_{i=1}^{n} a_i v_{ij}^2 \right) = \sum_{i=1}^{n} a_i \left(\sum_{j=1}^{m} v_{ij}^2 \right).$$

Now let $s_i = \sum_{j=1}^{m} v_{ij}^2$. Conversely, there is an $m \in \mathbb{N}$ such that $s_i = \sum_{j=1}^{m} v_{ij}^2$ for some $v_{ij} \in F$. Thus $m\rho$ is isotropic.

q.e.d.

(2.8) PROPOSITION If F is not formally real, then each quadratic form is weakly isotropic.

Proof: By (1.8)(c) there are $m \geq 1$ and $v_1,\ldots,v_n \in \dot{F}$ such that $\sum_{j=1}^{m} v_j^2 = 0$. The quadratic form $\langle a_1,\ldots,a_n \rangle$ is weakly isotropic, since $\sum_{j=1}^{m} (\sum_{i=1}^{n} a_i v_j^2) = 0$.

q.e.d.

As above, let P be a semiordering of F . Let ρ be a regular quadratic form of dimension n . We call

ρ <u>positive definite</u> w.r.t. P , if $\mathrm{sgn}_P\, \rho = n$

ρ <u>negative definite</u> w.r.t. P , if $\mathrm{sgn}_P\, \rho = -n$

ρ <u>definite</u> w.r.t. P , if $|\mathrm{sgn}_P\, \rho| = n$

ρ <u>indefinite</u> w.r.t. P , if $|\mathrm{sgn}_P\, \rho| < n$.

Obviously, if $\rho \simeq \langle a_1,\dots,a_n\rangle$, then by Theorem (2.6)

ρ positive definite w.r.t. P iff all $a_i \in P$

ρ negative definite w.r.t. P iff all $a_i \in -P$

ρ definite w.r.t. P iff (all $a_i \in P$ or all $a_i \in -P$)

ρ indefinite w.r.t. P iff there are i,j with $a_i, -a_j \in P$.

Now the argument in (2.7) shows that every quadratic form which is definite with respect to some semiordering of F is strongly anisotropic. The converse also holds.

(2.9) THEOREM <u>Let</u> ρ <u>be a regular quadratic form over</u> F . <u>Then</u> ρ <u>is weakly isotropic in</u> F <u>iff</u> ρ <u>is indefinite with respect to all semiorderings of</u> F .

<u>Proof</u>: If F is not formally real, the theorem follows from (2.8) and (1.15). Also "$\Rightarrow$" is trivial.

Let F be formally real and $\rho \simeq \langle a_1,\dots,a_n\rangle$. Assume ρ to be strongly anisotropic. Consider

$$P_0 = \{\Sigma\, a_i s_i \mid s_i \in S_F\} .$$

Obviously, $P_0 + P_0 \subset P_0$, and $F^2\, P_0 \subset P_0$. But also $P_0 \cap -P_0 = \{0\}$, since $\Sigma\, a_i s_i \in -P_0$ implies $\Sigma\, a_i s_i = -\Sigma\, a_i s_i'$ for some $s_i' \in S_F$.

But then $\Sigma a_i(s_i+s_i') = 0$, and the assumption implies $s_i + s_i' = 0$; hence $s_i = s_i' = 0$. Therefore P_o is a pre-semicone. By (1.13) , there is a semicone P such that $P_o \subset P$ or $-P_o \subset P$. Since $a_1,\ldots,a_n \in P_o$, the form ρ is definite with respect to P .

q.e.d.

Now let us introduce the principle:

WH : Every (regular) quadratic from over F is weakly isotropic if it is indefinite with respect to all orderings of F .

This principle will be called "Weak Hasse Principle".

(2.10) COROLLARY WH holds for every algebraic number field.

Proof: The corollary follows from (1.21) and (2.9).

This corollary justifies the name of the above principle, since it is a "weakened" form of the following statement:

(2.11) Every (regular) quadratic form of dimension ≥ 5 over an algebraic number field F is isotropic if it is indefinite with respect to all orderings of F .

This statement is a corollary of the famous "Local-Global-Principle" of Hasse and Minkowski for quadratic forms over algebraic number fields (compare [MH]).

(2.12) THEOREM [4] For a field F , (a) to (c) are equivalent:

(a) WH holds in F .

(b) Every semiordering of F is an ordering.

(c) $\langle 1,a,b,-ab\rangle$ is weakly isotropic in F for all $a,b \in \dot{F}$.

---

4) First proved in [P].

<u>Proof</u>: (b) ⇒ (a): trivial by (2.9).

(a) ⇒ (c): since $\langle 1,a,b,-ab\rangle$ is obviously indefinite with respect to every ordering of F.

(c) ⇒ (b): Suppose P is a proper semiordering of F. Then there are $a,b \in \dot{P}$ such that $ab \in -P$. Hence $1,a,b,-ab \in \dot{P}$. Therefore $\langle 1,a,b,-ab\rangle$ is not weakly isotropic.

q.e.d.

This theorem shows the importance of semiorderings of a field. But it does not really give a practical criterion to determine fields which satisfy WH. In general, one does not know the semiorderings of a field very well. In §9 we will give a much more practical criterion, which is based on the real places of a field (see also §6).

§3. REAL ALGEBRAIC CLOSURES

Let us recall the definition of a maximally ordered field given at the end of §1. A field F together with an ordering $\leq$ is called maximally ordered iff there is no proper algebraic extension field F_1 which admits an ordering $\leq_1$ extending the ordering $\leq$ of F.

(3.1) PROPOSITION Every positive element of a maximally ordered field F is a square. Hence F has a unique ordering.

Proof: Let P be the considered ordering of F. If $a \in P$, then P extends to $F(\sqrt{a})$ by (1.26)(a). Hence $\sqrt{a} \in F$, since F maximally ordered. Therefore $P \subset S_F$. But then $P = S_F$, which implies the uniqueness of P, by (1.11).

q.e.d.

A field F is called real closed if F is formally real, but has no formally real proper algebraic extension field F_1.

(3.2) LEMMA A field F is real closed iff it has a unique ordering and is maximally ordered.

Proof: "$\Rightarrow$" Let P be some ordering of F. As in the proof of (3.1), we get $P = S_F$. Hence F has a unique ordering and is obviously maximally ordered.

"$\Leftarrow$" Since F has a unique ordering P, any ordering P_1 of a formally real extension field F_1 extends P. Hence $F_1 = F$, if F_1 is algebraic over F.

q.e.d.

(3.3) THEOREM (Artin-Schreier [AS]) For a field F, (a), (b) and (c) are equivalent.

(a) F is real closed.

(b) F^2 is a positive cone of F and every polynomial of odd degree has a root in F.

(c) $F(\sqrt{-1})$ is algebraically closed and $F \neq F(\sqrt{-1})$.

Proof: (a) $\Rightarrow$ (b): This follows from (3.2), (3.1) and (1.26)(b).
(b) $\Rightarrow$ (c): Since $-1 \notin F^2$, we get $F(\sqrt{-1}) \neq F$ and char $F = 0$. Suppose there is a proper algebraic extension F' of $F(\sqrt{-1})$. Let F' be Galois over F and G its Galois group over F. The fixed field of a 2-Sylow subgroup of G is then a finite extension of F of odd degree. By (b), this degree can only be 1. Hence G is a 2-group. Now consider the Galois group G_1 of F' over $F(\sqrt{-1})$, which is again a 2-group. If G_1 is not trivial, it has a subgroup of index 2. The fixed field of this subgroup is a quadratic extension of $F(\sqrt{-1})$. This is impossible, since $F = F^2 \cup -F^2$ implies that every element of $F(\sqrt{-1})$ is a square.

(c) $\Rightarrow$ (a): First let us show $F^2 + F^2 \subset F^2$. Since $F(\sqrt{-1})$ is algebraically closed, for all $a,b \in F$, there are $c,d \in F$ such that $a+b\sqrt{-1} = (c+d\sqrt{-1})^2$. Hence we get $a = c^2-d^2$ and $b = 2cd$. This implies $a^2+b^2 = (c^2+d^2)^2 \in F^2$. Thus we get $S_F = F^2$. Now $F \neq F(\sqrt{-1})$ implies $-1 \notin S_F$, i.e. F is formally real. The only algebraic extension $F(\sqrt{-1})$ of F is not formally real; hence F is real closed.

q.e.d.

(3.4) LEMMA Let F be real closed and $\leq$ its unique ordering. For every $f \in F[x]$,

(a) f splits into irreducible factors of the type $x-a$ or $(x-a)^2 + b^2$ for some $a,b \in F$, and

(b) if $a < b$ and $f(a) < 0 < f(b)$, then there is a $c \in F$ such that $a < c < b$ and $f(c) = 0$.

Proof: (a): Since $F(\sqrt{-1})$ is algebraically closed by (3.3), f splits into irreducible factors of degree 1 or 2, i.e. of the type $x-a$ or $x^2-2ax + c = (x-a)^2 + (c-a^2)$. The second one can only be irreducible if $a^2-c \in -F^2$. But then $c-a^2 \in F^2$.

(b): Split f into irreducible factors as described in (a). Then a change of the sign of f has to come from some linear factor. Hence a zero $c \in F$ of f between a and b is easily found.

q.e.d.

Now let $< F,P >$ be an ordered field, i.e. F is a field and P is an ordering of F . We use the obvious notation $<F,P> \subset <F_1,P_1>$, if F_1 is an extension field of F together with an ordering P_1 extending P .

We call a field R a <u>real algebraic closure</u> of $<F,P>$ if

(3.5) (1) R is real closed.

(2) R is algebraic over F , and

(3) $P \subset S_R$.

(3.6) LEMMA <u>Every ordered field</u> $<F,P>$ <u>admits a real algebraic closure</u>.

<u>Proof</u>: By Zorn's Lemma in the set of algebraic order-extension fields of $<F,P>$, contained in some fixed algebraic closure of F , there is a maximal element. By (3.2) such a maximally ordered field is real closed.

q.e.d.

It is now natural to ask whether two real closures of $< F,P >$ are isomorphic. Artin and Schreier proved that the answer is yes.

<u>Remark</u>: Let $\varphi:R_1 \to R_2$ be a (field) isomorphism of two real closed fields. Then φ is also an order isomorphism. Since any positive a of R_1 equals b^2 for some $b \in R_1$, we get $\varphi(a) = \varphi(b)^2 \in R_2^2$. Hence $\varphi(a)$ is also positive in R_2 .

The proof of Artin and Schreier depends on the fact that the existence of a zero of $f \in F[x]$ in a real algebraic closure of F

can already be expressed in F itself. This was usually done using a theorem of Sturm. Quite recently Knebusch [Kn] gave a new proof using a certain quadratic form over F and Theorem (2.6). Becker and Spitzlay, in [BS], showed the connection with Sturm's theorem. In the following, we will present the proof of [BS].

Let $\langle F,P\rangle$ be an ordered field and $f \in F[x]$ an irreducible non-constant polynomial. Let $\alpha_1,\ldots,\alpha_n$ be all zeros of f in some algebraic closure $\tilde{F}$ of F. We define

$$\sigma_i := \sum_{r=1}^{n} \alpha_r^i \qquad (i \in \mathbb{N}) .$$

All σ_i are symmetric functions in $\alpha_1,\ldots,\alpha_n$ over $\mathbb{Z}$. Hence σ_i are rational functions of the coefficients of f; in particular $\sigma_i \in F$ for all i. Thus

$$\rho_f(X_1,\ldots,X_n) := \sum_{1\le r,s\le n} \sigma_{r+s-2} X_r X_s$$

is a quadratic form over F.

(3.7) THEOREM <u>For every real algebraic closure</u> R <u>of</u> $\langle F,P\rangle$ <u>in</u> $\tilde{F}$,

$$\mathrm{sgn}_P \rho_f = \text{number of } \alpha_i \in R .$$

<u>Proof</u>: Let $\beta_1,\ldots,\beta_m$ be all zeros of f in R. Since f is irreducible and $\mathrm{char}\, F = 0$, the claim of the theorem is $m = \mathrm{sgn}_P \rho_f$.

Now let $\gamma_1,\ldots,\gamma_\ell,\bar{\gamma}_1,\ldots,\bar{\gamma}_\ell$ be the remaining zeros of f in $\tilde{F}$. Since $\tilde{F} = R(\sqrt{-1})$, by (3.3) any element of $\tilde{F}$ has the form $a+b\cdot\sqrt{-1}$ with $a,b \in R$. As usual let $i = \sqrt{-1}$ and $\overline{a+bi} := a-bi$. By (3.4)(a), every zero γ of f which is not in R is a zero of some irreducible polynomial $(x-a)^2 + b^2 \in R[x]$. Hence $\bar{\gamma}$ is also a zero of f.

Now

$$\rho_f(X_1,\ldots,X_n) = \sum_{r,s,t} \alpha_t^{r-1+s-1} X_r X_s$$

$$= \sum_{t=1}^{n} \left(\sum_{r=1}^{n} \alpha_t^{r-1} X_r \right)^2$$

$$= \sum_{t=1}^{m} \left(\sum_{r=1}^{n} \beta_t^{r-1} X_r \right)^2 + \sum_{s=1}^{\ell} \left[\left(\sum_{r=1}^{n} \gamma_s^{r-1} X_r \right)^2 + \right.$$

$$\left. + \left(\sum_{r=1}^{n} \bar{\gamma}_s^{r-1} X_r \right)^2 \right]$$

$$= \sum_{t=1}^{m} Y_t^2 + \sum_{s=1}^{\ell} 2 \left[\left(\sum_{r=1}^{n} \frac{\gamma_s^{r-1} + \overline{\gamma_s^{r-1}}}{2} X_r \right)^2 + \right.$$

$$\left. + \left(\sum_{r=1}^{n} \frac{\gamma_s^{r-1} - \overline{\gamma_s^{r-1}}}{2} X_r \right)^2 \right]$$

$$= \sum_{t=1}^{m} Y_t^2 + \sum_{s=1}^{\ell} 2 \left(Y_{m+2s-1}^2 - Y_{m+2s}^2 \right)$$

using the substitution

$$Y_t = \sum_{r=1}^{n} \beta_t^{r-1} X_r \qquad (1 \leq t \leq m) ,$$

$$Y_{m+2s-1} = \sum_{r=1}^{n} \frac{\gamma_s^{r-1} + \overline{\gamma_s^{r-1}}}{2} X_r \, (1 \leq s \leq \ell) , \text{ and}$$

$$Y_{m+2s} = \sum_{r=1}^{n} \frac{\gamma_s^{r-1} - \overline{\gamma_s^{r-1}}}{2i} X_r \, (1 \leq s \leq \ell) .$$

This substitution is obviously over R. Its determinant is the Vandermonde determinante, which is different from zero since all α_i are different. The substitution shows that the quadratic form ρ_f is over R equivalent to $\rho' = m\langle 1\rangle \perp \ell\langle 2\rangle \perp \ell\langle -2\rangle$. The signature of ρ' with respect to R^2 is m. Hence, by (2.6), the signature of ρ_f is also m. Since R^2 extends P, we get $\mathrm{sgn}_P \rho_f = m$.

q.e.d.

(3.8) LEMMA <u>Let</u> R_1, R_2 <u>be real algebraic closures of the ordered field</u> $\langle F,P\rangle$ <u>and</u> $\langle F,P\rangle \subset \langle F_1,P_1\rangle \subset \langle R_1,R_1^2\rangle$ s.t. $[F_1:F]$ <u>is finite. Then there is an embedding of</u> $\langle F_1,P_1\rangle$ <u>into</u> $\langle R_2,R_2^2\rangle$ <u>which is the identity on</u> F.

Proof: Without loss of generality let R_1 , R_2 be contained in some algebraic closure $\tilde{F}$ of F . Let $f \in F[x]$ be the minimal polynomial of some $\alpha \in R_1$ such that $F_1 = F(\alpha)$. Hence, $\operatorname{sgn}_P \rho_f \neq 0$. But then, by Theorem (3.7) , f also has a root in R_2 . Thus there is an F-embedding (i.e. an embedding which is the identity on F) of F_1 into R_2 . Let $\sigma_1,\dots,\sigma_n$ be all f-embeddings of F_1 into R_2 , and suppose none of them is order-preserving. Then there are $a_1,\dots,a_n \in P_1$ such that $\sigma_i(a_i) \notin R_2^2$. Consider $F_2 = F_1(\sqrt{a_1},\dots,\sqrt{a_n}) \subset R_1$. Since again $[F_2:F]$ is finite, we may apply the same argument for F_2 as above for F_1 to get an F-embedding σ of F_2 into R_2 . Now $\sigma|_{F_1} = \sigma_i$ for some $1 \leq i \leq n$. But then $\sigma_i(a_i) = \sigma(a_i) = \sigma(\sqrt{a_i})^2 \in R_2^2$ gives a contradiction. Hence, one of the σ_i's embeds $\langle F_1,P_1\rangle$ into $\langle R_2,R_2^2\rangle$.

q.e.d.

This lemma is easily extended to the following

(3.9) LEMMA *Let* σ *be an order-isomorphism from* $\langle F_1,P_1\rangle$ *to* $\langle F_2,P_2\rangle$, R_i *a real algebraic closure of* $\langle F_i,P_i\rangle$ $(i=1,2)$ *and* $\langle F_1,P_1\rangle \subset \langle F_1',P_1'\rangle \subset \langle R_1,R_1^2\rangle$ *such that* $[F_1':F_1]$ *is finite. Then there is an extension of* σ *to an order-isomorphism from* $\langle F_1',P_1'\rangle$ *into* $\langle R_2,R_2^2\rangle$.

(3.10) THEOREM (Artin-Schreier [AS]) *Any ordered field* $\langle F,P\rangle$ *has a unique (up to isomorphism) real algebraic closure*.

Proof: The existence was proved in (3.6). Let R_1 , R_2 be two real algebraic closures of $\langle F,P\rangle$. Consider the set of order-isomorphisms

$$\sigma: \langle F_1,P_1\rangle \to \langle F_2,P_2\rangle$$

where $\langle F,P\rangle \subset \langle F_i,P_i\rangle \subset \langle R_i,R_i^2\rangle$ $(i=1,2)$ together with the partial ordering $\sigma_1 \subset \sigma_2$ (i.e. σ_2 extends σ_1). By Zorn's lemma there is a maximal isomorphism σ^* in this set. Using Lemma (3.9) for σ^* and σ^{*-1} , we conclude that σ^* is an isomorphism of $\langle R_1,R_1^2\rangle$ onto $\langle R_2,R_2^2\rangle$.

q.e.d.

From this fundamental theorem, we obtain some important corollaries.

(3.11) COROLLARY Let R be a real algebraic closure of $\langle F,P\rangle$. Let $F \subset F_1, F_2 \subset R$ be subfields of R and σ an isomorphism from F_1 onto F_2 such that $\sigma|_F = id$. Then σ is an order-isomorphism of $\langle F_1, R^2 \cap F_1\rangle$ and $\langle F_2, R^2 \cap F_2\rangle$ if and only if $\sigma = id$. In particular, $Aut(R/F) = \{id\}$.

Proof: By Theorem (3.10) (and its proof), σ extends to an automorphism σ^* of $\langle R,R^2\rangle$. We claim $\sigma^* = id$. Every $\alpha \in R$ is a zero of some polynomial $f \in F[X]$. Since $\sigma|_F = id$, $\sigma(\alpha)$ is also a zero of f. Hence, σ can only permute the zeros of f. But σ is order-preserving. Thus this permutation has to be the identity.

q.e.d.

(3.12) COROLLARY Let R be a real algebraic closure of $\langle F,P\rangle$ and $F \subset F(\alpha) \subset R$. Then the number of orderings of $F(\alpha)$ extending P equals the number of embeddings of $F(\alpha)$ into R which are the identity on F.

Proof: Any embedding $\sigma: F(\alpha) \to R$ such that $\sigma|_F = id$ yields an ordering $\sigma^{-1}(R^2)$ of $F(\alpha)$ extending P. Different embeddings σ_1 and σ_2 yield different orderings. Since $\sigma_1^{-1}(R^2) = \sigma_2^{-1}(R^2)$ implies $\sigma_2\sigma_1^{-1}$ is an order-isomorphism ($\neq id$) of $\langle\sigma_1(F(\alpha)), \sigma_1(F(\alpha)) \cap R^2\rangle$ and $\langle\sigma_2(F(\alpha)), \sigma_2(F(\alpha)) \cap R^2\rangle$, we have a contradiction to (3.11). Hence, the number of embeddings is less than or equal to the number of orderings. To get the equality, we have to show that every ordering P_1 of $F(\alpha)$ extending P equals $\sigma^{-1}(R^2)$ for some F-embedding σ of $F(\alpha)$ into R. Thus, let R_1 be a real algebraic closure of $\langle F(\alpha), P_1\rangle$. Since R and R_1 are both real algebraic closures of $\langle F,P\rangle$, by (3.10), there is an isomorphism σ of R_1 and R such that $\sigma|_F = id$. But then $\sigma_1 := \sigma|_{F(\alpha)}$ is an F-embedding of $F(\alpha)$ into R such that $\sigma_1^{-1}(R^2) = P_1$.

q.e.d.

Theorem (3.3), for instance, shows that $\mathbb{R}$ is real closed. Moreover, it shows that every real closed field behaves very similar to the reals $\mathbb{R}$ concerning algebraic properties, e.g. (3.3)(b). But these are only special cases of a very general principle, called the Tarski-Principle. It states that any "elementary" property of $\mathbb{R}$ also holds in every real closed field. This principle will be proved in the next paragraphs (also "elementary" will be defined there). For this proof, we need two more algebraic lemmas.

(3.13) LEMMA Let R be real closed and F a subfield of R. Then the algebraic closure $\tilde{F}^R$ of F in R is again real closed.

Proof: $\tilde{F}^R$ is the set of all $a \in R$ which are algebraic over F. $P := \tilde{F}^R \cap R^2$ is a positive cone of $\tilde{F}^R$. Since for every $a \in P$ the polynomial x^2-a has a zero in R, it also has one in $\tilde{F}^R$. Therefore the squares of $\tilde{F}^R$ form a positive cone. Now let $f \in \tilde{F}^R[X]$ be a polynomial of odd degree. Since f has a zero in R, it also has one in $\tilde{F}^R$. Thus, (3.3) shows that $\tilde{F}^R$ is real closed.

q.e.d.

(3.14) LEMMA Let $\langle F, \leq \rangle$ be an ordered field and R_1, R_2 two real closed fields extending $\langle F, \leq \rangle$. For every finite number of polynomials $f_1, \ldots, f_r$, $g_1, \ldots, g_s \in F[X]$, there is some $a \in R_1$ which satisfies

$$(*) \qquad f_1(a) = 0, \ldots, f_r(a) = 0, \; g_1(a) > 0, \ldots, g_s(a) > 0$$

if and only if there is such an a in R_2. (The set of f_i's or the set of g_i's may be empty).

Proof: Let F_1 and F_2 be the algebraic closure of F in R_1 and R_2, resp. By (3.13) F_1 and F_2 are real closed. Hence, by (3.10) there is an isomorphism σ such that

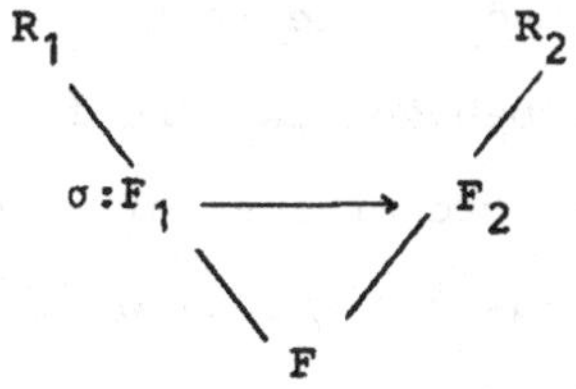

Let $a \in R_1$ satisfy (*). To get an element of R_2 also satisfying (*), let us go via F_1 and F_2. This means we claim that there is a $b \in F_1$ satisfying (*). But then $\sigma(b)$ also satisfies (*) and is an element of R_2. To prove the claim, we first assume that there is at least one non-constant f_i. But then $f_i(a) = 0$ implies $a \in F_1$. Next, we assume there is no non-constant f_i. Thus we have to find a $b \in F_1$ satisfying $g_i(b) > 0$ for all $1 \leq i \leq s$. Every g_i has only a finite number of zeros in R_1 (which are also in F_1). Let A be the union of all these zeros. If $A = \emptyset$, then $g_i(0) > 0$, for all $1 \leq i \leq s$. Since otherwise, some g_j would have a zero in R_1 by (3.4)(b). If $A \neq \emptyset$, we distinguish three cases.

<u>1. case</u>: $a > \max A$. Then $g_i(\max A+1) > 0$, for all $1 \leq i \leq s$. This follows again from (3.4)(b).

<u>2. case</u>: $a < \min A$. Then $g_i(\min A-1) > 0$ for all $1 \leq i \leq s$.

<u>3. case</u>: a lies between two elements of A. Let $\alpha < a < \beta$ for some $\alpha, \beta \in A$ such that no other element of A lies between α and β. But then $g_i(\frac{\alpha+\beta}{2}) > 0$ for all $1 \leq i \leq s$, again by (3.4)(b).

q.e.d.

§ 4. SOME NOTIONS FROM MODEL THEORY

As mentioned at the end of the last paragraph up to a certain extent real closed fields behave like the field $\mathbb{R}$ of reals. We will make this precise now. The properties carrying over from $\mathbb{R}$ to arbitrary real closed fields are socalled "elementary" properties. This notion as well as others used in the present paragraph are from model theory. Hence we will give first a short introduction to the basic notions and results of model theory. The more interested reader is referred e.g. to [B SL] or [Ch K].

In model theory one usually considers mathematical "structures" like the field $\mathbb{Q}$ of rational numbers and "elementary" properties like " $\leq$ is a linear ordering" or " any positive element is a square". The crucial feature of an elementary property is that it can be described using the quantification "for all" and "there is" only for elements of the considered structure and never for subsets of this structure. For instance, the property "every bounded non-empty subset admits a least upper bound" will not be an elementary property of the field of real numbers.

To characterize these elementary properties one usually introduces a "formal language" which then is used to "describe" these properties. We carry out all this only for fields and ordered fields. It can be done much more generally. The interested reader is referred to [B SL] and [Ch K] once more.

The <u>elementary language of fields</u> is based on the symbols

$$+,\ -,\ \cdot,\ 0,\ 1,\ x_1,\ x_2,\ x_3,\ \ldots$$

together with the logical connectives

$$\sim,\ \wedge,\ \vee,\ \rightarrow,\ \leftrightarrow,\ \forall,\ \exists,\ =$$

and the brackets) and (.

A structure or realization of this language is a 6-tuple

$$\mathfrak{A} = \langle |\mathfrak{A}|, +^{\mathfrak{A}}, -^{\mathfrak{A}}, \cdot^{\mathfrak{A}}, 0^{\mathfrak{A}}, 1^{\mathfrak{A}} \rangle$$

where $A = |\mathfrak{A}|$ is a non-empty set, $0^{\mathfrak{A}}$ and $1^{\mathfrak{A}}$ are elements of A and

$$+^{\mathfrak{A}} : A \times A \to A$$
$$-^{\mathfrak{A}} : A \to A$$
$$\cdot^{\mathfrak{A}} : A \times A \to A \, .$$

$+^{\mathfrak{A}}, -^{\mathfrak{A}}, \cdot^{\mathfrak{A}}, 0^{\mathfrak{A}}, 1^{\mathfrak{A}}$ are called the interpretations of the formal signs $+, -, \cdot, 0, 1$ in $\mathfrak{A}$. The variables $x_1, x_2, \ldots$ are supposed to vary over the elements of A. As an example of a structure we may consider the set $\mathbb{Q}$ of rationals together with the usual interpretations of $+, -, \cdot, 0, 1$. But also the set $\mathbb{Z}$ of integers together with the usual interpretations form a structure of the elementary language of fields.

The logical connectives are interpreted as follows

$\sim$	$\wedge$	$\vee$	$\to$	$\leftrightarrow$	$\forall$	$\exists$	$=$
not	and	or	if	iff	for all	there is	equal

Instead of "if" and "iff" we already used "$\Rightarrow$" and "$\Leftrightarrow$" resp.

We will consider the elementary language of ordered fields as well. To get this we simply add the sign $<$ to the symbols of the language of fields. A structure of the elementary language of ordered fields then is a 7-tuple

$$\mathfrak{A} = \langle |\mathfrak{A}|, +^{\mathfrak{A}}, -^{\mathfrak{A}}, \cdot^{\mathfrak{A}}, 0^{\mathfrak{A}}, 1^{\mathfrak{A}}, <^{\mathfrak{A}} \rangle$$

such that the first 6 components form a structure of the elementary language of fields and $<^{\mathfrak{A}}$ is a binary relation on $A = |\mathfrak{A}|$.

To both languages we will also consider certain extensions. They are given by adding a finite number of <u>new constants</u> $c_1,\dots,c_n$ to the symbols of the language. A structure of this enlarged language now looks like

$$<\mathfrak{A},\ a_1,\dots,a_n>$$

where $\mathfrak{A}$ is a structure of the original language and $a_1,\dots,a_n \in |\mathfrak{A}|$ are the interpretations of $c_1,\dots,c_n$ in $\mathfrak{A}$ resp.

For the elementary language of fields, enlarged by the constants $c_1,\dots,c_n$, we define the set of <u>terms</u> Tm_n recursively by

(1) $0,1,\ c_1,\dots,c_n,\ x_1,x_2,\dots \in Tm_n$

(2) if $t_1,t_2 \in Tm_n$, then also (t_1+t_2), $(t_1 \cdot t_2)$, $-t_1 \in Tm_n$.

The case $n = 0$ means that no constant c_1 is present. The set of <u>formulas</u> Fml_n is defined recursively by

(1) $t_1 = t_2 \in Fml_n$ for all $t_1,t_2 \in Tm_n$

(2) if $\varphi,\psi \in Fml_n$ and $x \in Vbl$, then also $\sim\varphi$, $(\varphi \wedge \psi)$, $(\varphi \vee \psi)$, $(\varphi \rightarrow \psi)$, $(\varphi \leftrightarrow \psi)$, $\forall x\ \varphi$, $\exists x\ \varphi \in Fml_n$,

where $Vbl = \{x_1,x_2,\dots\}$. The formulas $t_1 = t_2$ are called <u>prime formulas</u>.

For the elementary language of ordered fields we define

$$Tm_n^< := Tm_n$$

and $Fml_n^<$ similar to Fml_n by

(1) $t_1 = t_2,\ t_1 < t_2 \in Fml_n^<$

(2) as above .

Hence $Fml_n \subset Fml_n^<$. Again the formulas from (1) are called prime formulas.

We call $t \in Tm_n$ a constant term, if no variable $x \in Vbl$ occurs in t. In a structure $\mathfrak{A}$ the interpretation $t^{\mathfrak{A}}$ of a constant term t is obviously determined by the interpretations $0^{\mathfrak{A}}, 1^{\mathfrak{A}}, c_1^{\mathfrak{A}}, \ldots, c_n^{\mathfrak{A}}$ of the symbols $0, 1, c_1, \ldots, c_n$, if we define

$$(t_1 + t_2)^{\mathfrak{A}} = t_1^{\mathfrak{A}} +^{\mathfrak{A}} t_2^{\mathfrak{A}}, \quad (-t_1)^{\mathfrak{A}} = -^{\mathfrak{A}} t_1^{\mathfrak{A}}$$

and

$$(t_1 \cdot t_2)^{\mathfrak{A}} = t_1^{\mathfrak{A}} \cdot^{\mathfrak{A}} t_2^{\mathfrak{A}} .$$

If t_1 and t_2 are constant terms and $\mathfrak{A} = \langle \ldots, <^{\mathfrak{A}} \rangle$ is a structure of the elementary language of ordered fields (e.g. $\mathbb{R}$), then we will say that the formula $t_1 < t_2$ "holds" in $\mathfrak{A}$ iff $t_1^{\mathfrak{A}} <^{\mathfrak{A}} t_2^{\mathfrak{A}}$. But it is meaningless to ask whether the formula $x_1 < 1$ holds in $\mathbb{R}$ or not. The reason for this is that x_1 occurs "free" in $x_1 < 1$. This means that it depends on the value of x_1 whether $x_1 < 1$ holds in $\mathbb{R}$ or not. However, in the formula $\forall x_1 \; x_1 < 1$, x_1 no longer occurs free. It is "bound" by $\forall x_1$. In this case it makes sense to say that this formula does not hold in $\mathbb{R}$.

We now define the set $Fr(\varphi)$ of free variables in the formula φ recursively by

$$Fr(t_1 = t_2) = \{x \in Vbl \mid x \text{ occurs in } t_1 \text{ or in } t_2\}$$
$$Fr(t_1 < t_2) = \{x \in Vbl \mid x \text{ occurs in } t_1 \text{ or in } t_2\}$$
$$Fr(\sim\varphi) = Fr(\varphi)$$
$$Fr(\varphi \wedge \psi) = Fr(\varphi \vee \psi) = Fr(\varphi \rightarrow \psi) = Fr(\varphi \leftrightarrow \psi) = Fr(\varphi) \cup Fr(\psi)$$
$$Fr(\forall x\, \varphi) = Fr(\exists x\, \varphi) = Fr(\varphi) \setminus \{x\}.$$

The elements of the sets

$$Sent_n := \{\varphi \in Fml_n \mid Fr(\varphi) = \emptyset\}$$
$$Sent_n^< := \{\varphi \in Fml_n^< \mid Fr(\varphi) = \emptyset\}$$

are called <u>sentences</u> of the corresponding language.

We are now in the position to formalize the intuitive meaning of "a sentence φ holds in a structure $\mathfrak{A}$" by a recursive definition. Let $\varphi \in \mathrm{Sent}_n^<$ and $\mathfrak{B} = <\mathfrak{A}, a_1, \ldots, a_n>$. We then define the relation $\mathfrak{B} \models \varphi$ which reads as "φ <u>holds in</u> $\mathfrak{B}$" by

$$\begin{array}{lll}
\mathfrak{B} \models (t_1 = t_2) & \text{iff} & t_1^{\mathfrak{B}} = t_2^{\mathfrak{B}} \\
\mathfrak{B} \models (t_1 < t_2) & \text{iff} & t_1^{\mathfrak{B}} <^{\mathfrak{B}} t_2^{\mathfrak{B}} \\
\mathfrak{B} \models (\sim \varphi) & \text{iff} & \text{not } \mathfrak{B} \models \varphi \\
\mathfrak{B} \models (\varphi \wedge \psi) & \text{iff} & (\mathfrak{B} \models \varphi \text{ and } \mathfrak{B} \models \psi) \\
\mathfrak{B} \models (\varphi \vee \psi) & \text{iff} & (\mathfrak{B} \models \varphi \text{ or } \mathfrak{B} \models \psi) \\
\mathfrak{B} \models (\varphi \rightarrow \psi) & \text{iff} & (\mathfrak{B} \models \varphi \text{ implies } \mathfrak{B} \models \psi) \\
\mathfrak{B} \models (\varphi \leftrightarrow \psi) & \text{iff} & (\mathfrak{B} \models \varphi \text{ iff } \mathfrak{B} \models \psi) \\
\mathfrak{B} \models \forall x\, \varphi(x) & \text{iff} & \text{for all } a \in |\mathfrak{A}| : <\mathfrak{B}, a> \models \varphi(c_{n+1}) \\
\mathfrak{B} \models \exists x\, \varphi(x) & \text{iff} & \text{there is } a \in |\mathfrak{A}| : <\mathfrak{B}, a> \models \varphi(c_{n+1}) .
\end{array}$$

If for a formula φ we write $\varphi(x_1, \ldots, x_m)$, we indicate by this that $\mathrm{Fr}(\varphi) \subset \{x_1, \ldots, x_m\}$, i.e. there are no other variables free in φ than at most $x_1, \ldots, x_n$. Let us also remark that we did not fix the number n of constants in the above definition.

By C_n and $C_n^<$ let us denote the class of structures $<\mathfrak{A}, a_1, \ldots, a_n>$ of the elementary language of fields and of ordered fields resp. enlarged by the constants $c_1, \ldots, c_n$. For the following we fix one of these classes and call it simply C. By Sent we denote the set of sentences of the language corresponding to C.

For any subset Σ of Sent, $\mathfrak{A} \in C$ is called a <u>model</u> of Σ if $\mathfrak{A} \models \sigma$ for all $\sigma \in \Sigma$. The class of all models of Σ

$$\mathrm{mod}(\Sigma) = \{\mathfrak{A} \in C \mid \mathfrak{A} \models \sigma \text{ for all } \sigma \in \Sigma\}$$

is called the <u>modelclass of</u> Σ. An arbitrary subclass K of C is called a modelclass or elementary class or axiomatizable if it is the modelclass of some $\Sigma \subset \text{Sent}$. Σ then is called an axiom system of K. If Σ can be chosen as a finite set, K is also called finitely axiomatizable.

Now we are able to give a precise definition of an "elementary property" of a structure. Set theoretically a property of a considered object is a certain class of objects ("the collection of all objects with this property") to which this object belongs. Thus by an elementary property we mean an elementary subclass K of C . We say $\mathfrak{A}$ has this property if $\mathfrak{A} \in K$.

(4.1) EXAMPLES

(1) Subclasses of C_o

(i) The class of fields is elementary since the usual field axioms are from Sent_o . Consider e.g. the distributivity

$$\forall x_1 \ \forall x_2 \ \forall x_3 \ \ x_1 \cdot (x_2 + x_3) = (x_1 \cdot x_2) + (x_1 \cdot x_3) \ .$$

(ii) The class of algebraically closed fields is elementary. To see this we add to the field axioms others stating that every non-constant polynomial has a zero. Thus for every $n \geq 1$ we add the axiom[5)]

$$\forall x_1, \dots, x_n \ \exists y \quad y^n + x_n y^{n-1} + \dots + x_1 = 0 \ .$$

In contrast to (i), (ii) is axiomatized by an infinite number of sentences. Moreover, (ii) is not finitely axiomatizable.

(iii) The class of formally real fields is elementary. To (i) we add for every $n \geq 1$ the axiom

$$\forall x_1, \dots, x_n \quad x_1^2 + \dots + x_n^2 \neq -1 \ .$$

---

5) Mostly we use the common abbreviations in writing formal sentences.

By (1.8) this is an axiom system for the class of formally real fields.

(iv) The class of <u>real closed fields</u> is elementary. To give an axiom system we use Theorem (3.3). To (i) we add for every $n \geq 1$

$$\forall x_1,\ldots,x_{2n-1}\ \exists y \quad y^{2n-1} + x_{2n-1}\, y^{2n-2} + \ldots + x_1 = 0 .$$

These axioms state that any polynomial of odd degree has a zero. Furthermore we add the axioms

$$\forall x_1, x_2 \ \exists y \quad x_1^2 + x_2^2 = y^2$$

$$\forall x_1, x_2 \quad (x_1^2 = -x_2^2 \rightarrow x_1 = x_2 = 0)$$

$$\forall x_1 \ \exists y \quad (x_1 = y^2 \vee -x_1 = y^2) \ ;$$

these axioms state that the set of squares forms a positive cone.

(2) <u>Subclasses of</u> $C_o^<$

(i) The class of <u>ordered fields</u> is elementary. We add the axioms (1.1) (i)-(vi) to the field axioms.

(ii) The class of <u>maximally ordered fields</u> is elementary. To (i) we first add the axioms

$$\forall x_1 \ \exists y \quad (0 < x_1 \rightarrow x_1 = y^2)$$

stating that the squares form a positive cone. Hence in any model there is a unique ordering. But then by Lemma (3.2) it suffices to express that the considered model is real closed. Therefore we add all the axioms from (1) (iv) stating that any odd-degree polynomial has a zero [6].

---

6) We get another axiom system simply by joining the axioms of (1)(iv) and (2)(i).

(iii) The class of <u>archimedean ordered fields</u> is <u>not</u> an elementary class. The reason for this fact is that we cannot define the subset $\mathbb{N}$ of an ordered field by a formula of the elementary language of ordered fields. One way to define $\mathbb{N}$ in an ordered field F is by the intersection of all inductive subsets A of F. By $\mathrm{Ind}(A)$ we mean

$$0 \in A \wedge \forall x (x \in A \rightarrow x+1 \in A);$$

then

$$y \in \mathbb{N} :\longleftrightarrow \forall A (\mathrm{Ind}(A) \rightarrow y \in A).$$

Now

$$\forall x \ \exists y \ (y \in \mathbb{N} \wedge x < y)$$

expresses that the ordering is archimedean. But this "axiom" uses a quantification $\forall A$ over all subsets of the considered field. As we will see later there is no way to replace this by other sentences which are elementary.

(3) <u>Subclasses of</u> C_{n+1} <u>and</u> $C^{<}_{n+1}$

The field axioms (1)(i) together with

$$\exists y \quad c_{n+1} \, y^n + \ldots + c_1 = 0$$

or

$$\exists y \quad c_{n+1} \, y^n + \ldots + c_1 > 0$$

axiomatize the subclass of structures $\langle \mathfrak{A}, a_1, \ldots, a_{n+1} \rangle$ such that $\mathfrak{A}$ is a field or an ordered field resp. and the polynomial

$$f(X) = a_{n+1} \, X^n + \ldots + a_1$$

has a zero in $|\mathfrak{A}|$ or a positive value resp. .

Now we will introduce a <u>topology on the</u> fixed <u>class</u> C. As a "base of open sets" we use the classes

$$\mathrm{Mod}(\sigma) = \{\mathfrak{A} \in C \mid \mathfrak{A} \models \sigma\}$$

for $\sigma \in \text{Sent}$. These classes form a base since

$$\text{Mod}(\sigma_1) \cap \text{Mod}(\sigma_2) = \text{Mod}(\sigma_1 \wedge \sigma_2) .$$

From

$$C \setminus \text{Mod}(\sigma) = \text{Mod}(\sim \sigma)$$

it follows that $\text{Mod}(\sigma)$ is open and closed.

(4.2) LEMMA <u>A subclass of</u> C <u>is closed iff it is an elementary class</u>.

<u>Proof</u>: Since the classes $\text{Mod}(\sigma)$ are open and closed, every closed subset K of C equals $\bigcap_{\sigma\in\Sigma} \text{Mod}(\sigma)$ for some $\Sigma \subset \text{Sent}$. But

$$\bigcap_{\sigma\in\Sigma} \text{Mod}(\sigma) = \{\mathfrak{A} \in C \mid \mathfrak{A} \models \sigma \quad \text{for all} \quad \sigma \in \Sigma\} = \text{Mod}(\Sigma) .$$

Hence K is elementary and conversely.

q.e.d.

<u>Remark</u>: In dealing with proper classes we may have difficulties with the usual axioms of set theory. But there are several ways to avoid this. One way is to extend the axioms of set theory suitably. Another way is to consider only a certain subset of C containing enough structures. We will not go into this further.

The most fundamental theorem from model theory is the

(4.3) COMPACTNESS THEOREM C <u>is (quasi-)compact, i.e. every open covering of</u> C <u>contains a finite subcover</u>.

We will not prove this theorem here. The interested reader is referred to [B SL] or [Ch K]. It is easy to see that C is not a Hausdorff space, even if we restrict ourselves to isomorphism classes of C . An equivalent version of (4.3) is:

(4.3') COMPACTNESS THEOREM <u>There is a model of an axiom system</u> $\Sigma \subset \text{Sent}$ <u>if every finite subset of</u> Σ <u>admits a model</u>.

<u>Proof</u>: ((4.3) ⇒ (4.3')) Suppose there is no model of Σ . But then

$$\emptyset = \mathrm{Mod}(\Sigma) = \bigcap_{\sigma\in\Sigma} \mathrm{Mod}(\sigma) .$$

Hence $\{\mathrm{Mod}(\sim \sigma) \mid \sigma \in \Sigma\}$ is an open covering of C . Thus there is a finite subset $\Sigma_0 \subset \Sigma$ such that $\{\mathrm{Mod}(\sim \sigma) \mid \sigma \in \Sigma_0\}$ already covers C . But then

$$\emptyset = \bigcap_{\sigma\in\Sigma_0} \mathrm{Mod}(\sigma) = \mathrm{Mod}(\Sigma_0) .$$

Therefore the finite subset Σ_0 of Σ admits no model.

q.e.d.

As a first application we prove

(4.4) THEOREM <u>The class of archimedean ordered fields is not elementary</u>

<u>Proof</u>: Here we let $C = C_0^<$. Suppose there is a subset $\Sigma \subset \mathrm{Sent}_0^<$ such that

$$\mathrm{Mod}(\Sigma) = \text{class of archimedean ordered fields.}$$

Now add one constant c_1 to the language and consider the following subset of $\mathrm{Sent}_1^<$:

$$\Sigma_1 := \Sigma \cup \{n\cdot 1 < c_1 \mid n \in \mathbb{N}\}$$

where $n\cdot 1$ is the constant term $1 + \ldots + 1$ (n-times). Every finite subset Σ_0 of Σ_1 is contained in some set

$$\Sigma_0' = \Sigma \cup \{n\cdot 1 < c_1 \mid 0 \leq n \leq m\} .$$

Obviously the real numbers together with the interpretation of c_1 by $m+1$ form a model of Σ_0' .

Applying now the Compactness Theorem (4.3') to $\mathrm{Sent} = \mathrm{Sent}_1^<$ we get a model $\langle \mathfrak{A}, a_1\rangle$ of Σ_1 . Note that Σ does not contain the constant c_1 . Hence we have $\mathfrak{A} \in \mathrm{Mod}(\Sigma)$. By the assumption on Σ ,

$\mathfrak{A}$ then is an archimedean ordered field. On the other hand $<\mathfrak{A},a_1> \in \mathrm{Mod}(\Sigma_1)$ and hence $<\mathfrak{A},a_1> \models (n\cdot 1 < c_1)$ for all $n \in \mathbb{N}$. This implies $n <^{\mathfrak{A}} a_1$ for all $n \in \mathbb{N}$. But this is impossible since the ordering $<^{\mathfrak{A}}$ is archimedean. q.e.d.

As before let C denote C_n or $C_n^<$ for some $n \in \mathbb{N}$. Sent again denotes the corresponding set Sent_n or $\mathrm{Sent}_n^<$. Let $\Gamma \subset \mathrm{Sent}$ be closed under $\wedge,\vee$ and $\sim$, i.e.

$$\varphi,\psi \in \Gamma \Rightarrow (\varphi \wedge \psi),(\varphi \vee \psi), \sim\varphi \in \Gamma \quad .$$

Two structures $\mathfrak{A}, \mathfrak{A}' \in C$ are called <u>Γ-equivalent</u> ($\mathfrak{A} \equiv_\Gamma \mathfrak{A}'$) if for all $\gamma \in \Gamma$

$$\mathfrak{A} \models \gamma \iff \mathfrak{A}' \models \gamma \ .$$

This means, there is no possibility of distinguishing $\mathfrak{A}$ from $\mathfrak{A}'$ by sentences from Γ. To say it positively, this means that every "property" $\gamma \in \Gamma$ carries over from $\mathfrak{A}$ to $\mathfrak{A}'$ and conversely (precisely: $\mathfrak{A} \in \mathrm{Mod}(\gamma) \iff \mathfrak{A}' \in \mathrm{Mod}(\gamma)$). If $\Gamma = \mathrm{Sent}$ we call $\mathfrak{A}$ and $\mathfrak{A}'$ <u>elementarily equivalent</u> and write $\mathfrak{A} \equiv \mathfrak{A}'$.

(4.5) THEOREM <u>Let</u> K,K' <u>be two elementary subclasses of</u> C. <u>If no</u> $\mathfrak{A} \in K$ <u>and</u> $\mathfrak{A}' \in K'$ <u>are</u> Γ<u>-equivalent, then there is a</u> $\gamma \in \Gamma$ <u>separating</u> K <u>from</u> K', i.e. $K \subset \mathrm{Mod}(\gamma)$ <u>and</u> $K' \subset \mathrm{Mod}(\sim\gamma)$.

<u>Proof</u>: Any two points $\mathfrak{A} \in K$ and $\mathfrak{A}' \in K'$ can be separated by some $\gamma \in \Gamma$ since $\mathfrak{A}$ is not Γ-equivalent to $\mathfrak{A}'$.

Fix $\mathfrak{A} \in K$ and choose a separating class $\mathrm{Mod}(\gamma)$ to any $\mathfrak{A}' \in K'$, i.e. $\mathfrak{A} \in \mathrm{Mod}(\gamma)$ and $\mathfrak{A}' \in \mathrm{Mod}(\sim\gamma)$. The classes $\mathrm{Mod}(\sim\gamma)$ then cover K'. Since K' is elementary it is a closed subclass of the compact space C. Thus there is a finite cover $\mathrm{Mod}(\sim\gamma_1) \cup \ldots \cup \mathrm{Mod}(\sim\gamma_m) = \mathrm{Mod}(\sim(\gamma_1 \wedge \ldots \wedge \gamma_m))$ of K'. Since $\mathfrak{A} \in \mathrm{Mod}(\gamma_1) \cap \ldots \cap \mathrm{Mod}(\gamma_m) = \mathrm{Mod}(\gamma_1 \wedge \ldots \wedge \gamma_m)$ we obtained a class $\mathrm{Mod}(\gamma)$ separating $\mathfrak{A}$ from K', i.e. $\mathfrak{A} \in \mathrm{Mod}(\gamma)$ and $K' \subset \mathrm{Mod}(\sim\gamma)$.

Now choose to every $\mathfrak{A} \in K$ a class $\mathrm{Mod}(\gamma)$ separating $\mathfrak{A}$ from K'. These classes cover K. Again by the compactness of C there is a finite subcover $\mathrm{Mod}(\gamma_1') \cup \ldots \cup \mathrm{Mod}(\gamma_r') = \mathrm{Mod}(\gamma_1' \vee \ldots \vee \gamma_r')$. Since $K' \subset \mathrm{Mod}(\sim\gamma_1') \cap \ldots \cap \mathrm{Mod}(\sim\gamma_r') = \mathrm{Mod}(\sim(\gamma_1' \vee \ldots \vee \gamma_r'))$ we obtained a separating class for K and K'.

q.e.d.

The most interesting class $\Gamma \subset$ Sent consists of the <u>quantifier free</u> sentences, i.e. sentences φ such that the symbols $\forall$ and $\exists$ do not occur in φ. Obviously this class is closed under $\wedge, \vee$ and $\sim$. If Sent = Sent_n we denote the class of quantifier free sentences by Γ_n; in case Sent = $\mathrm{Sent}_n^<$ we denote it by $\Gamma_n^<$. Now let K be a subclass of C_o (or $C_o^<$). We say that K <u>admits elimination of quantifiers</u> if for every $\varphi \in \mathrm{Sent}_n$ (or $\varphi \in \mathrm{Sent}_n^<$) there is a $\gamma \in \Gamma_n$ (or $\gamma \in \Gamma_n^<$) such that

$$<\mathfrak{A}, a_1,\ldots,a_n> \models (\varphi \leftrightarrow \gamma)$$

for all $\mathfrak{A} \in K$ and $a_1,\ldots,a_n \in |\mathfrak{A}|$. In this case we say φ <u>is equivalent to</u> γ <u>mod</u> K.

In the next paragraph we will prove that the class of maximally ordered fields admits elimination of quantifiers. From this fact the "Tarski-Principle" will follow easily. To make this deduction we will use the following notion.

Let $\mathfrak{A}$ and $\mathfrak{A}'$ be structures of C_o or of $C_o^<$ resp. . Then $\mathfrak{A}$ is called a <u>substructure</u> of $\mathfrak{A}'$ if $|\mathfrak{A}| \subset |\mathfrak{A}'|$, $0^{\mathfrak{A}} = 0^{\mathfrak{A}'}$, $1^{\mathfrak{A}} = 1^{\mathfrak{A}'}$ and the operations of $\mathfrak{A}$ (and the relation $<^{\mathfrak{A}}$, if present) are restrictions of the corresponding ones of $\mathfrak{A}'$. In this case we write $\mathfrak{A} \subset \mathfrak{A}'$. We call $\mathfrak{A}$ an <u>elementary substructure</u> of $\mathfrak{A}'$ if $\mathfrak{A} \subset \mathfrak{A}'$ and

$$<\mathfrak{A}, a_1,\ldots,a_n> \equiv <\mathfrak{A}', a_1,\ldots,a_n>

for all $n \in \mathbb{N}$ and all $a_1,\dots,a_n \in |\mathfrak{A}|$. In this case we write $\mathfrak{A} \prec \mathfrak{A}'$.

(4.6) LEMMA <u>If the class</u> K <u>admits elimination of quantifiers, then for all</u> $\mathfrak{A},\mathfrak{A}' \in K$:

$$\mathfrak{A} \subset \mathfrak{A}' \Rightarrow \mathfrak{A} \prec \mathfrak{A}' .$$

<u>Proof</u>: Let $\varphi \in \mathrm{Sent}_n^{(<)}$. Choose some $\gamma \in \Gamma_n^{(<)}$ such that

$$(*) \qquad \langle \mathfrak{B}, b_1,\dots,b_n\rangle \models (\varphi \leftrightarrow \gamma)$$

for all $\mathfrak{B} \in K$ and $b_1,\dots,b_n \in |\mathfrak{B}|$. Since γ is without quantifiers, obviously

$$\langle \mathfrak{A}, a_1,\dots,a_n\rangle \models \gamma \Longleftrightarrow \langle \mathfrak{A}', a_1,\dots,a_n\rangle \models \gamma$$

for all $a_1,\dots,a_n \in |\mathfrak{A}|$. But then by (*)

$$\langle \mathfrak{A}, a_1,\dots,a_n\rangle \models \varphi \Longleftrightarrow \langle \mathfrak{A}', a_1,\dots,a_n\rangle \models \varphi$$

for all $a_1,\dots,a_n \in |\mathfrak{A}|$.

q.e.d.

The following theorem is a criterion for quantifier elimination.

(4.7) THEOREM <u>An elementary class</u> $K \subset C_o^{(<)}$ <u>admits elimination of quantifiers iff for any two structures</u> $\mathfrak{A},\mathfrak{A}' \in K$ <u>and all</u> $a_1,\dots,a_n \in |\mathfrak{A}|$, $a_1',\dots,a_n' \in |\mathfrak{A}'|$:

$$\langle \mathfrak{A}, a_1,\dots,a_n\rangle \equiv_{\Gamma_n^{(<)}} \langle \mathfrak{A}', a_1',\dots,a_n'\rangle$$

<u>implies</u>

$$\langle \mathfrak{A}, a_1,\dots,a_n\rangle \models \exists x\, \varphi \Longleftrightarrow \langle \mathfrak{A}', a_1',\dots,a_n'\rangle \models \exists x\, \varphi$$

<u>for all</u> $\exists x\, \varphi \in \mathrm{Sent}_n^{(<)}$ <u>such that</u> φ <u>is without quantifiers</u>.

<u>Proof</u>: The proof of the non-trivial direction proceeds in two steps.

Step 1: Let $\exists x \varphi \in Sent_n^{(<)}$ such that φ is without quantifiers. We will show that $\exists x \varphi$ is equivalent to some $\gamma \in \Gamma_n^{(<)}$ mod K .

By assumption $K = Mod(\Sigma)$ for some $\Sigma \subset Sent_0^{(<)}$. Consider the closed subclasses of $C_n^{(<)}$

$$K_1 := Mod(\Sigma \cup \{\exists x \varphi\})$$
$$K_2 := Mod(\Sigma \cup \{\sim \exists x \varphi\}) .$$

By assumption of the theorem there are no structures $<\mathfrak{A}_1, a_1,\ldots,a_n> \in K_1$ and $<\mathfrak{A}', a_1',\ldots,a_n'> \in K_2$ which are $\Gamma_n^{(<)}$-equivalent. Hence by Theorem (4.5) there is $\gamma \in \Gamma_n^{(<)}$ such that $K_1 \subset Mod(\gamma)$ and $K_2 \subset Mod(\sim\gamma)$. From this it follows that

$$<\mathfrak{A}, a_1,\ldots,a_n> \models (\exists x \varphi \rightarrow \gamma)$$
$$<\mathfrak{A}, a_1,\ldots,a_n> \models (\sim \exists x \varphi \rightarrow \sim\gamma)$$

and hence

$$<\mathfrak{A}, a_1,\ldots,a_n> \models (\exists x \varphi \leftrightarrow \gamma)$$

for all $\mathfrak{A} \in K$ and $a_1,\ldots,a_n \in |\mathfrak{A}|$. Therefore $\exists x \varphi$ is equivalent to γ mod K .

Step 2: Now we prove the assertion of the theorem by induction on the length of the considered sentence.

If a sentence ψ is without quantifiers we may take ψ as some γ . If ψ equals $(\psi_1 \wedge \psi_2)$, by induction hypothesis ψ_1 and ψ_2 are equivalent to some quantifier free γ_1 and γ_2 resp. . But then γ is equivalent to $(\gamma_1 \wedge \gamma_2)$ mod K . By the same argument we treat the connectives $\sim$, $\vee$, $\rightarrow$, $\leftrightarrow$. Now let $\psi \in Sent_n^{(<)}$ equal $\exists x \varphi(x)$. Since $\varphi(c_{n+1})$ is shorter than ψ , by induction hypothesis there is $\gamma'(c_{n+1}) \in \Gamma_{n+1}^{(<)}$ such that

$$<\mathfrak{A}, a_1,\ldots,a_{n+1}> \models (\varphi(c_{n+1}) \leftrightarrow \gamma'(c_{n+1}))$$

for all $\mathfrak{A} \in K$ and $a_1,\ldots,a_{n+1} \in |\mathfrak{A}|$. But then also

$$<\mathfrak{A}, a_1,\ldots,a_n> \models (\exists x\, \varphi(x) \leftrightarrow \exists x\, \gamma'(x)).$$

Now by Step 1 there is $\gamma \in \Gamma_n^{(<)}$ such that

$$<\mathfrak{A}, a_1,\ldots,a_n> \models (\exists x\, \gamma'(x) \leftrightarrow \gamma).$$

Thus

$$<\mathfrak{A}, a_1,\ldots,a_n> \models (\exists x \varphi \leftrightarrow \gamma)$$

for all $\mathfrak{A} \in K$ and $a_1,\ldots,a_n \in |\mathfrak{A}|$. The final case $\forall x\, \varphi$ is treated by replacing $\forall x\, \varphi$ by $\sim \exists x \sim \varphi$ which is logically equivalent.

q.e.d.

§ 5. THE TRANSFER-PRINCIPLE FOR REAL CLOSED FIELDS

The Transfer Principle

This principle will be a corollary to the following theorem.

(5.1) THEOREM (Tarski [T]) The class of maximally ordered fields admits elimination of quantifiers.

Proof: To prove this theorem we will use the criterion (4.7). Consider the class K consisting of maximally ordered fields $\mathfrak{A} = \langle R, < \rangle$ [7]. By Example (4.1)(2)(ii), K is an elementary class. Now assume

(*) $\quad \langle R, <, a_1, \ldots, a_n \rangle \equiv_{\Gamma_n^<} \langle R', <', a_1', \ldots, a_n' \rangle$

where $\langle R, < \rangle$ and $\langle R', <' \rangle$ are maximally ordered fields and $a_1, \ldots, a_n \in R$, $a_1', \ldots, a_n' \in R'$. Consider

$$F := \mathbb{Q}(a_1, \ldots, a_n) \subset R$$
$$F' := \mathbb{Q}(a_1', \ldots, a_n') \subset R'$$

together with the induced orderings, denoted again by $<$ and $<'$ resp. We claim

$$\langle F, < \rangle \simeq \langle F', <' \rangle .$$

This is a consequence of (*). An arbitrary element of F looks like

$$\frac{f(a_1, \ldots, a_n)}{g(a_1, \ldots, a_n)} \quad \text{where} \quad f, g \in \mathbb{Z}[X_1, \ldots, X_n] .$$

The isomorphism may be defined by

---

7) From now on we do not mention the field operations following the usage of §1-§3 .

$$\frac{f(a_1,\dots,a_n)}{g(a_1,\dots,a_n)} \mapsto \frac{f(a_1',\dots,a_n')}{g(a_1',\dots,a_n')}$$

To show that this is an isomorphism it suffices to prove

$$h(a_1,\dots,a_n) = 0 \iff h(a_1',\dots,a_n') = 0$$

for all $h \in \mathbb{Z}[X_1,\dots,X_n]$. But this immediately follows from (*) since the sentence $h(c_1,\dots,c_n) = 0$ contains no quantifier. Similarly it follows that

$$h(a_1,\dots,a_n) > 0 \iff h(a_1',\dots,a_n') >' 0 .$$

Now let us identify the isomorphic subfields $\langle F,<\rangle$ and $\langle F',<'\rangle$. Hence

$$\begin{array}{ccc} \langle R,<,a_1,\dots,a_n\rangle & & \langle R',<',a_1,\dots,a_n\rangle \\ & \diagdown \quad \diagup & \\ & \langle F,<,a_1,\dots,a_n\rangle & \end{array}$$

Consider a sentence $\exists x\, \varphi(x) \in \mathrm{Sent}_n^<$ where $\varphi(x)$ is quantifier free. We assume $\langle R,<,a_1,\dots,a_n\rangle \models \exists x\, \varphi(x)$. We have to show $\langle R',<',a_1,\dots,a_n\rangle \models \exists x\, \varphi(x)$. (The converse follows by symmetry.) To prove this we will use Lemma (3.14). For this reason we first transform $\varphi(x)$ equivalently into a certain normal form. Since $\varphi(x)$ is without quantifiers it is built by use of $\sim,\wedge,\vee,\rightarrow,\leftrightarrow$ from some prime formulas $f_i' = f_i''$ and $g_j' < g_j''$. Here f_i', f_i'', g_j', g_j'' are terms built from the constants $0,1,c_1,\dots,c_n$ and the variable x . We replace these prime formulas equivalently by $f_i(x) = 0$ and $g_j(x) > 0$ with suitable terms f_i and g_j . Next we eliminate $\rightarrow$, $\leftrightarrow$ and $\sim$. This can be done replacing

$(\psi_1 \leftrightarrow \psi_2)$ by $(\psi_1 \rightarrow \psi_2) \wedge (\psi_2 \rightarrow \psi_1)$

$(\psi_1 \rightarrow \psi_2)$ by $\sim(\psi_1 \wedge \sim \psi_2)$

$\sim\sim\psi$ by ψ

$\sim(\psi_1 \wedge \psi_2)$ by $\sim\psi_1 \vee \sim\psi_2$

$$\sim(\psi_1 \vee \psi_2) \text{ by } \sim\psi_1 \wedge\sim \psi_2$$

(Now $\sim$ only occurs in front of prime formulas!)

$$\sim t = 0 \text{ by } t > 0 \vee t < 0$$
$$\sim t > 0 \text{ by } t = 0 \vee t < 0 .$$

Now $\varphi(x)$ is built only by use of $\wedge$ and $\vee$ from prime formulas. Replacing $\psi_1 \wedge (\psi_2 \vee \psi_3)$ by $(\psi_1 \wedge \psi_2) \vee (\psi_1 \wedge \psi_3)$ we can express $\varphi(x)$ as

$$\psi_1(x) \vee\ldots\vee \psi_m(x)$$

where every $\psi_i(x)$ is a conjunction of prime formulas.

From the assumption $\langle R,<,a_1,\ldots,a_n\rangle \models \exists x\, \varphi(x)$ it follows that $\langle R,<,a_1,\ldots,a_n\rangle \models \exists x\, \psi_i(x)$ for some $i \in \{1,\ldots,m\}$. But now $\psi_i(x)$ looks like

$$f_1(x) = 0 \wedge\ldots\wedge f_r(x) = 0 \wedge g_1(x) > 0 \wedge\ldots\wedge g_s(x) > 0 .$$

Hence from Lemma (3.14) we deduce that also $\langle R',<',a_1,\ldots,a_n\rangle \models$ $\models \exists x\, \psi_i(x)$. This implies trivially $\langle R',<',a_1,\ldots,a_n\rangle \models \exists x\, \varphi(x)$.

q.e.d.

(5.2) COROLLARY <u>Let</u> R <u>and</u> R' <u>be real closed fields. Then</u> $R \subset R'$ <u>implies</u> $\langle R,<\rangle \prec \langle R',<'\rangle$ <u>where</u> $<$ <u>and</u> $<'$ <u>are the unique orderings of</u> R <u>and</u> R' <u>resp</u>.

<u>Proof</u>: Obviously $R \subset R'$ implies $\langle R,<\rangle \subset \langle R',<'\rangle$. Now (5.2) follows from (3.2), (5.1) and (4.6).

q.e.d.

In our applications we will use one more corollary. The next one is usually called the "Tarski-Principle".

(5.3) COROLLARY <u>Let</u> R <u>and</u> R' <u>be real closed fields and</u> $<,<'$ <u>denote their unique orderings. Then</u> $\langle R,<\rangle \equiv \langle R',<'\rangle$.

Proof: Let $\bar{\mathbb{Q}}$ and $\bar{\mathbb{Q}}'$ be the algebraic closure of the prime field $\mathbb{Q}$ in R and R' resp. By (3.13) both are real closed and hence isomorphic by (3.10) (since $\mathbb{Q}$ has a unique ordering). Identifying $\bar{\mathbb{Q}}$ with $\bar{\mathbb{Q}}'$ we get the situation

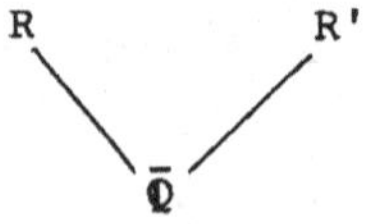

Now (5.2) implies

$$\langle R,<\rangle \succ \langle \bar{\mathbb{Q}},<\rangle \prec \langle R',<'\rangle.$$

By the Definition of $\prec$, in particular we get

$$\langle R,<\rangle \equiv \langle \bar{\mathbb{Q}},<\rangle \equiv \langle R',<'\rangle$$

and hence $\langle R,<\rangle \equiv \langle R',<'\rangle$.

q.e.d.

Remark: The class of real closed fields does not admit elimination of quantifiers. (Note that the corresponding elementary language does not contain the symbol $<$.) To show this, consider the field $F = Q(\sqrt{2})$. Since F has two embeddings into the real algebraic closure of $\mathbb{Q}$ it has two orderings $<_1$ and $<_2$. Without loss of generality we may assume $0 <_1 \sqrt{2}$ and $\sqrt{2} <_2 0$. Consider the real algebraic closures R_1 and R_2 of $\langle F,<_1\rangle$ and $\langle F,<_2\rangle$ resp. Let φ be the sentence $\exists x\ c_1 = x^2$. Then we get $\langle R_1,\sqrt{2}\rangle \models \varphi$ and $\langle R_2,\sqrt{2}\rangle \models \sim\varphi$. Now suppose that φ is equivalent to some γ without quantifiers. From $\langle R_1,\sqrt{2}\rangle \models \gamma$ we deduce $\langle F,\sqrt{2}\rangle \models \gamma$ since γ contains no quantifier. But $\langle F,\sqrt{2}\rangle \models \gamma$ then implies $\langle R_2,\sqrt{2}\rangle \models \gamma$. Since γ is equivalent to φ in all real closed fields we would get the contradiction $\langle R_2,\sqrt{2}\rangle \models \varphi$.

Let us now indicate how to get the analogues to (5.1), (5.2) and (5.3) for algebraically closed fields.

(5.4) THEOREM <u>The class of algebraically closed fields admits elimination of quantifiers.</u>

<u>Proof:</u> By Example (4.1)(1)(ii) the class K of algebraically closed fields is elementary. We use again the criterion (4.7). Hence we assume

$$\langle F, a_1, \ldots, a_n \rangle \equiv_{\Gamma_n} \langle F', a_1', \ldots, a_n' \rangle$$

where $F, F' \in K$ and $a_1, \ldots, a_n \in F$, $a_1', \ldots, a_n' \in F'$. Furthermore let $\exists x\, \varphi \in \mathrm{Sent}_n$ such that φ is quantifier free. Now the proof runs along the lines of that from (5.1). There is only one difference. Prime formulas of the type $t_1 < t_2$ do not occur and $\sim f_i = 0$ cannot be expressed positively. Hence eventually $\varphi(x)$ is equivalent to some $\psi_1(x) \vee \ldots \vee \psi_m(x)$ where any $\psi_i(x)$ looks like

$$f_1(x) = 0 \wedge \ldots \wedge f_r(x) = 0 \wedge \sim g_1(x) = 0 \wedge \ldots \wedge \sim g_s(x) = 0 .$$

To continue the proof similarly to (5.1) we now need a lemma similar to (3.14) for algebraically closed fields, where $g_j(a) > 0$ is replaced by $g_j(a) \neq 0$. But this version of the lemma is trivial to prove.

q.e.d.

(5.5) COROLLARY <u>Let</u> F <u>and</u> F' <u>be algebraically closed fields. Then</u> $F \subset F'$ <u>implies</u> $F \prec F'$.

<u>Proof:</u> follows from (5.4) and (4.6).

(5.6) COROLLARY (Lefschetz-Principle) <u>Let</u> F <u>and</u> F' <u>be algebraically closed and of the same characteristic. Then</u> $F \equiv F'$.

<u>Proof:</u> similar to that of (5.3). The identical characteristic is needed to get isomorphic prime fields.

q.e.d.

Some Applications

First we prove a theorem from Artin which is an answer to Hilbert's 17th problem.

(5.7) THEOREM (Artin [A]) Let R be real closed, $f \in R[X_1,\dots,X_m]$ and $f(b_1,\dots,b_m) \geq 0$ for all $b_1,\dots,b_m \in R$. Then $f \in S_{R(X_1,\dots,X_m)}$ = set of sums of squares in $R(X_1,\dots,X_m)$.

Proof: Suppose $f \notin S_{R(X_1,\dots,X_m)}$. By Corollary (1.9) there is an ordering $<'$ of the field $R(X_1,\dots,X_m)$ such that $f <' 0$. Obviously $<'$ extends the unique ordering $<$ of R. Now let R' be the real algebraic closure of $\langle R(X_1,\dots,X_m),<'\rangle$. The extension of $<'$ to R' is again denoted by $<'$. From $R \subset R'$ by (5.2) we get $\langle R,<\rangle \prec \langle R',<'\rangle$. Let

$$f(X_1,\dots,X_m) = a_1\ h_1(X_1,\dots,X_m) +\dots+ a_n\ h_n(X_1,\dots,X_m)$$

where h_j are monic monomials and $a_j \in R$. Now $f <' 0$ implies

$$(R',<',a_1,\dots,a_n) \models \exists x_1,\dots,x_m \sum_{j=1}^{n} c_j\ h_j(x_1,\dots,x_m) < 0 .$$

From $\langle R,<\rangle \prec \langle R',<'\rangle$ we get in particular

$$\langle R,<,a_1,\dots,a_n\rangle \equiv \langle R',<',a_1,\dots,a_n\rangle$$

and hence

$$\langle R,<,a_1,\dots,a_n\rangle \models \exists x_1,\dots,x_m \sum_{j=1}^{m} c_j\ h_j(x_1,\dots,x_m) < 0 .$$

This is equivalent to: there are $b_1,\dots,b_m \in R$ such that $f(b_1,\dots,b_m) < 0$, contradicting the assumption of the theorem.

q.e.d.

Next we prove Hilbert's Nullstellensatz.

(5.8) THEOREM Let F be algebraically closed, $I \subset F[X_1,\dots,X_m]$ an ideal and $f \in F[X_1,\dots,X_m]$ such that each zero $(b_1,\dots,b_m) \in F^m$

of I (i.e. $g(b_1,\dots,b_m) = 0$ for all $g \in I$) is also a zero of f. Then f is in the radical of I (i.e. $f^r \in I$ for some $r \in \mathbb{N}$).

Proof: Suppose $f^{\mathbb{N}} \cap I = \emptyset$. Since $f^{\mathbb{N}}$ is closed under multiplication there is (by a theorem of Krull) a prime ideal $J \supset I$ such that $f^{\mathbb{N}} \cap J = \emptyset$. By Hilbert's Basissatz, I is finitely generated, say $I = (g_1,\dots,g_s)$. Now let F' denote the algebraic closure of the quotient field of the domain $F[X_1,\dots,X_m]/J$. If $y_1,\dots,y_m$ denote the images of $X_1,\dots,X_m$ in F' resp. we obtain $g_i(y_1,\dots,y_m) = 0$ for all $1 \le i \le s$ and $f(y_1,\dots,y_m) \neq 0$ (since $f \notin J$). As in the proof of (5.7) we replace all coefficients, say $a_1,\dots,a_n$, in f and also $g_1,\dots,g_s$ by constants $c_1,\dots,c_n$. By $f^c, g_1^c,\dots,g_s^c$ we denote the result of this replacement. Hence we get

$$\langle F',a_1,\dots,a_n\rangle \models \exists x_1,\dots,x_m(g_1^c(x_1,\dots,x_m) = 0 \wedge\dots\wedge g_s^c(x_1,\dots,x_m) = 0 \wedge \sim f^c(x_1,\dots,x_m) = 0) .$$

By (5.5), $F \subset F'$ implies $F \prec F'$. In particular we get

$$\langle F,a_1,\dots,a_n\rangle \equiv \langle F',a_1,\dots,a_n\rangle .$$

Hence

$$\langle F,a_1,\dots,a_n\rangle \models \exists x_1,\dots,x_m(g_1^c(x_1,\dots,x_m) = 0 \wedge\dots\wedge g_s^c(x_1,\dots,x_m) = 0 \wedge \sim f^c(x_1,\dots,x_m) = 0) .$$

But this is equivalent to: there are $b_1,\dots,b_m \in F$ such that $f(b_1,\dots,b_m) \neq 0$ and $g_i(b_1,\dots,b_m) = 0$ for all $1 \le i \le s$, which implies $g(b_1,\dots,b_m) = 0$ for all $g \in I$. This contradicts the assumption of the theorem.

q.e.d.

(5.9) COROLLARY Let F be algebraically closed and $I \subset F[X_1,\dots,X_m]$ an ideal. Then I has a zero in F iff $1 \notin I$.

Proof: If $1 \in I$, I obviously has no zero in F. If I has no zero in F, every zero of I is a zero of 1. Hence by (5.8)

$1^r \in I$ for some $r \in \mathbb{N}$.

q.e.d.

We shall now prove a theorem similar to (5.7) which also implies the so-called "Real Nullstellensatz". Thereby we will not restrict ourselves to real closed fields but will use arbitrary ordered fields. (5.7) (and analogously (5.8)) can be generalized similarly.

First let us generalize the notion S_F. If A is a ring and B a subset of A, we define

$$S_A(B) := \{\sum_{i=1}^{n} b_i\, a_i^2 \mid n \in \mathbb{N},\, b_i \in B,\, a_i \in A\}$$

and let $S_A := S_A(A^2)$. If A is a field, this notation coincides with the one introduced earlier.

(5.10) THEOREM Let $\langle F,P\rangle$ <u>be an ordered field and</u> $f \in F[X_1,\dots,X_m]$ <u>such that</u> $f(b_1,\dots,b_m) > 0$ <u>for all</u> $b_1,\dots,b_m \in \bar{F}$, <u>where</u> $\bar{F}$ <u>is the real algebraic closure of</u> $\langle F,P\rangle$ <u>and</u> $<$ <u>its unique ordering. Then there are</u> $s_1,s_2 \in S_{F[X_1,\dots,X_m]}(P)$ <u>such that</u> $f = (1+s_1)s_2^{-1}$.

<u>Proof</u>: Let $A := F[X_1,\dots,X_m]$ and suppose there are no $s_1,s_2 \in S_A(P)$ such that $f = (1+s_1)s_2^{-1}$. Hence $f\,S_A(P) \cap (1+S_A(P)) = \emptyset$. Now let $I := f\,S_A(P)$ and $P_0 := S_A(P) - I$. One easily checks that P_0 satisfies (1.3)(1)-(4). To show (3), let $-1 \in P_0$. But then $-1 = s_1 - f\,s_2$ for some $s_1,s_2 \in S_A(P)$, contradicting $f\,S_A(P) \cap (1+S_A(P)) = \emptyset$. Hence P_0 is a pre-positive cone of A. By Lemma (1.4) P_0 can be extended to some pre-positive cone P_1 of A such that $P_1 \cup -P_1 = A$ and $P_1 \cap -P_1 =: J$ is a prime ideal of A. P_1 yields a pre-positive cone P_2 of A/J such that $P_2 \cup -P_2 = A/J$ and $P_2 \cap -P_2 = \{0/J\}$. This holds, since $a/J = -b/J$ for some $a,b \in P_1$ implies $a+b \in J \subset -P_1$ and hence $a = (a+b) - b \in -P_1$, showing $a \in J$. Now an easy exercise shows that P_2 extends to a positive cone P' of the quotient field F' of A/J. Obviously

$\langle F,P\rangle \subset \langle F',P'\rangle$. By $\bar{F}'$ we denote as usual the real algebraic closure of $\langle F',P'\rangle$ and by $\bar{F}$ the real algebraic closure of $\langle F,P\rangle$ in $\bar{F}'$. Now by (5.2), $\bar{F} \subset \bar{F}'$ implies $\langle \bar{F},<\rangle \prec \langle \bar{F}',<'\rangle$, where $<$ and $<'$ denote the unique ordering of $\bar{F}$ and $\bar{F}'$ resp.

By our construction we know that $I \subset -P_1$. If $y_1,\dots,y_m$ are the images of $X_1,\dots,X_n$ in F', we therefore get $f(y_1,\dots,y_m) \le' 0$. Using now the same argument as in the proof of (5.7) (replacing R by $\bar{F}$, R' by $\bar{F}'$, and < 0 by ≤ 0) $\langle \bar{F},<\rangle \prec \langle \bar{F}',<'\rangle$ implies that there are $b_1,\dots,b_m \in \bar{F}$ such that $f(b_1,\dots,b_m) \le 0$. But this contradicts the assumption of the theorem.

q.e.d.

The next theorem may be called "Weak Real Nullstellensatz". It is the analogue to (5.9).

(5.11) THEOREM <u>Let</u> $\langle F,P\rangle$ <u>be an ordered field and</u> $I \subset F[X_1,\dots,X_m]$ <u>an ideal. Then</u> I <u>has a zero in the real algebraic closure</u> $\bar{F}$ <u>of</u> $\langle F,P\rangle$ <u>if and only if</u>

$$(1 + S_{F[X_1,\dots,X_m]}(P)) \cap I = \emptyset .$$

<u>Proof</u>: Obviously no element of $1 + S_{F[X_1,\dots,X_m]}(P)$ has a zero in $\bar{F}$. Hence the intersection of this set with I is empty, if I has a zero in $\bar{F}$.

Now assume I has no zero in $\bar{F}$. By Hilbert's Basissatz $I = (f_1,\dots,f_n)$ for some $f_1,\dots,f_n \in F[X_1,\dots,X_m]$. Since I has no zero in $\bar{F}$, also $\sum_{i=1}^{n} f_i^2$ has no zero in $\bar{F}$ (a zero of this sum is also a zero of every f_i and hence I). Let $f = \sum_{i=1}^{n} f_i^2$. From $f(b_1,\dots,b_m) > 0$ for all $b_1,\dots,b_m \in \bar{F}$ we get $f = (1+s_1)s_2^{-1}$ for some $s_1,s_2 \in S_{F[X_1,\dots,X_m]}(P)$ by Theorem (5.10). Hence $fs_2 \in (1+S_{F[X_1,\dots,X_m]}(P)) \cap I$.

q.e.d.

To get an analogue to (5.8), denote the subset $\mathcal{R}(I) = \{f \mid f^{2n} + s \in I$ for some $n \in \mathbb{N}$, $s \in S_{F[X_1,\dots,X_m]}(P)\}$ of $F[X_1,\dots,X_m]$ as real radical of the ideal $I \subset F[X_1,\dots,X_m]$. The following theorem may be called "Real Nullstellensatz".

(5.12) THEOREM (Dubois [8]) Let $\langle F,P \rangle$ be an ordered field, $I \subset F[X_1,\dots,X_m]$ an ideal and $f \in F[X_1,\dots,X_m]$ such that any zero of I in the real algebraic closure $\bar{F}$ of $\langle F,P \rangle$ is also a zero of f. Then f belongs to the real radical (I) of I.

Proof: For the proof we use a trick from Rabinowitsch. Consider the ring $A' = F[X_1,\dots,X_m,Y]$ and let $A = F[X_1,\dots,X_m]$. Since every zero $(b_1,\dots,b_m)$ of I in $\bar{F}$ is also a zero of f, the ideal $I' = IA' + (1-fY)A'$ of A' has no zero $(b_1,\dots,b_m,b)$ in $\bar{F}$. But then by (5.11)

$$(1 + S_{A'}(P)) \cap I' \neq \emptyset .$$

Hence

$$1 + \Sigma\, a_i\, g_i'^{\,2} = \Sigma\, f_j\, h_j' + (1-fY)h'$$

for some $f_j \in I$, $a_i \in P$ and h_j', h', $g_i' \in A'$. Substituting $\frac{1}{f}$ for Y and multiplying by some suitable power f^{2n} one gets $f^{2n} + s \in I$ and

$$s = f^{2n}\, \Sigma\, a_i\, g_i'(X_1,\dots,X_m,\tfrac{1}{f})^2 \in S_A(P) .$$

Thus $f \in \mathcal{R}(I)$.

q.e.d.

Remark: Since the converse of (5.12) is trivially true, this characterization of $\mathcal{R}(I)$ shows in particular that $\mathcal{R}(I)$ is indeed an ideal of $F[X_1,\dots,X_m]$.

---

8) Compare [Du] and [Du E]. The proof given here, especially Theorem (5.10), seems to be new.

We continue with one more application to real algebraic geometry.

Let $f \in F[X_1,X_2]$. By $D_i f$ we denote the polynomial derivative of f with respect to X_i $(i = 1,2)$. Then $D_1 f \neq 0$ and $D_2 f \neq 0$ express that f is neither constant in X_1 nor in X_2.

Note that every ordering of a field F induces a topology on F, generated by the usual open intervals.

(5.13) THEOREM Let R be real closed and f an irreducible polynomial of $R[X_1,X_2]$ such that $D_1 f \neq 0$ and $D_2 f \neq 0$. Assume that the function field of f over R is formally real. Then there are $\alpha,\beta \in R$ and neighborhoods U of α and V of β in R such that in $U \times V$ the equation $f(X_1,X_2) = 0$ is uniquely solvable (in X_1 as well as in X_2) and $D_1 f$ and $D_2 f$ don't vanish.

Proof: In case $R = \mathbb{R}$ and $D_1 f(\alpha,\beta) \neq 0$, $D_2 f(\alpha,\beta) \neq 0$ and $f(\alpha,\beta) = 0$ for some $\alpha,\beta \in \mathbb{R}$, the existence of $U,V \subset \mathbb{R}$ satisfying the statement of the theorem is well known.

The proof of the theorem proceeds in two steps. First we "transfer" this fact from $\mathbb{R}$ to R. Then we prove the existence of $\alpha,\beta \in R$ satisfying $D_1 f(\alpha,\beta) \neq 0$, $D_2 f(\alpha,\beta) \neq 0$ and $f(\alpha,\beta) = 0$.

In the first step we cannot use the polynomial f itself, since its coefficients are from R. Hence we replace the coefficients of f, say $a_1,\dots,a_n$, by variables $y_1,\dots,y_n$. Let f^y denote the result of the replacement. Now the following sentence ψ expresses in R with its unique ordering $<$ what we mentioned above.

$$\psi := \forall\, y_1 \dots y_n \;\forall\, x_1 x_2 [f^y(x_1,x_2) = 0 \wedge D_1 f^y(x_1,x_2) \neq 0$$
$$\wedge\, D_2 f^y(x_1,x_2) \neq 0 \rightarrow \varphi]$$

where φ expresses the statement of the theorem, i.e.

$$\varphi := \exists\, z_1 z_2 (z_1, z_2 > 0 \wedge \varphi_1 \wedge \varphi_2 \wedge \varphi_3)$$

$$\varphi_1 := \forall\, u[\,|u-x_1| < z_1 \rightarrow \overset{=1}{\exists}\, v(|v-x_2| < z_2 \wedge f^y(u,v) = 0)]$$

$$\varphi_2 := \forall\, v[\,|v-x_2| < z_2 \rightarrow \overset{=1}{\exists}\, u(|u-x_1| < z_1 \wedge f^y(u,v) = 0)]$$

$$\varphi_3 := \forall uv(|u-x_1| < z_1 \wedge |v-x_2| < z_2 \rightarrow$$

$$D_1 f^y(u,v) \neq 0 \wedge D_2 f^y(u,v) \neq 0)$$

Here $\overset{=1}{\exists}\, v$ is an abbreviation of the formula expressing "there is exactly one v ..." .

Since we know $\langle\mathbb{R},<\rangle \models \psi$, (5.3) implies $\langle R,<\rangle \models \psi$, where $<$ also denotes the unique ordering of R . Thus we shall have proved the theorem, once we have shown the existence of $\alpha,\beta \in R$ satisfying

$$(*) \qquad f(\alpha,\beta) = 0\ , \quad D_1 f(\alpha,\beta) \neq 0\ , \quad D_2 f(\alpha,\beta) \neq 0\ .$$

Let F be the function field of f . Then there are $x_1, x_2 \in F$ such that $F = R(x_1,x_2)$ and $f(x_1,x_2) = 0$. Since $f(x_1,\cdot)$ is also an irreducible polynomial in X_2 over $R(x_1)$, $D_2 f(x_1,x_2) \neq 0$. Similarly we also obtain $D_1 f(x_1,x_2) \neq 0$. Now let R_1 denote the real algebraic closure of F with respect to some ordering of F . Let f^c denote the result of f by replacing the coefficients this time by constants $c_1,\ldots,c_n$ of the formal language. Then obviously

$$\langle R_1,a_1,\ldots,a_n\rangle \models \exists\, x_1 x_2 (f^c(x_1,x_2) = 0 \wedge D_1 f^c(x_1,x_2) \neq 0 \wedge D_2 f^c(x_1,x_2) \neq 0)\ ,$$

But then by (5.2) the same sentence also holds in $\langle R,a_1,\ldots,a_n\rangle$, since $R \subset R_1$. This expresses the existence of $\alpha,\beta \in R$ satisfying (*).

An alternative proof of the second step makes use of the Real Nullstellensatz (5.11). Consider the ideal

$$I = (f(X_1,X_2)\ ,\ X_3 D_1 f(X_1,X_2) - 1\ ,\ X_4 D_2 f(X_1,X_2) - 1)$$

of the ring $R[X_1,\dots,X_4] =: A$. As we showed above, I has a zero in R_1 . But then $I \cap (1+S_A) = \emptyset$. Hence by (5.11) I has a zero in R , i.e. there are $\alpha,\beta,\gamma,\delta \in R$ such that $f(\alpha,\beta) = 0$, $\gamma D_1 f(\alpha,\beta) = 1$, $\delta D_2 f(\alpha,\beta) = 1$. This trivially implies (*).

q.e.d.

As a final application of Tarski's Principle let us prove that any universal (elementary) property of the ordered field $\langle\mathbb{R},<\rangle$ is also true in every other ordered field $\langle F,<\rangle$ (not necessarily real closed; compare the introduction). By a _universal elementary property of_ $\langle\mathbb{R},<\rangle$ we mean any $\varphi \in \mathrm{Sent}_0^<$ (= set of sentences of the elementary language of ordered fields) such that $\langle\mathbb{R},<\rangle \models \varphi$ and

$$\varphi = \forall\, x_1 \dots x_n \quad \psi\,(x_1,\dots,x_n)$$

where ψ is quantifier free. Any "rule" of dealing with field operations and the ordering is of this type, e.g.

(1.1)(v) $\qquad \forall\, x_1x_2x_3(x_1 \le x_2 \to x_1+x_3 \le x_2+x_3).$

(5.14) THEOREM _Any universal elementary property of_ $\langle\mathbb{R},<\rangle$ _also holds in any ordered field_.

Proof: Let $\langle F,<\rangle$ be an arbitrary ordered field. Let R denote its real algebraic closure. The unique ordering of R is also denoted by <. Now let φ be a universal property of $\langle\mathbb{R},<\rangle$. Then by (5.3), $\langle\mathbb{R},<\rangle \models \varphi$ implies $\langle R,<\rangle \models \varphi$. Obviously a _universal_ property true in $\langle R,<\rangle$ is also true in $\langle F,<\rangle$. Hence $\langle F,<\rangle \models \varphi$.

q.e.d.

This theorem shows that the concept of an ordered field is a very natural generalization of the field $\mathbb{R}$ of real numbers together with its usual "ordering".

§ 6. THE SPACE OF ORDERINGS AND SEMIORDERINGS

So far we always considered a field F together with a fixed ordering or a semiordering P. Now let us look at the set of all orderings and the set of all semiorderings of a given field F. Sometimes it is more convenient to exclude 0 from a positive cone $P \subset F$. If $P \subset F$ is a positive cone or a semicone resp. (i.e. it satisfies (1.2) (1)-(4), or (1)-(4) on page 5) then $\dot{P}$ obviously satisfies

$(\dot{1})$ $\dot{P} + \dot{P} \subset \dot{P}$

$(\dot{3})$ $\dot{P} \cap -\dot{P} = \emptyset$ [9]

$(\dot{4})$ $\dot{P} \cup -\dot{P} = \dot{F}$

and the corresponding second condition. Conversely if some subset $\dot{P}$ of $\dot{F}$ satisfies $(\dot{1})$, $(\dot{3})$, $(\dot{4})$ and the corresponding second condition, then $\dot{P} \cup \{0\} =: P$ obviously satisfies (1.2) (1)-(4), or (1)-(4) on page 5 resp. Hence, we will call a subset $P \subset \dot{F}$ satisfying $(\dot{1})$, $(\dot{3})$, $(\dot{4})$ and the corresponding second condition an ordering or a semiordering of F as well. Now let

$$X_F := \{P \subset \dot{F} \mid P \text{ an ordering of } F\}$$

$$Y_F := \{P \subset \dot{F} \mid P \text{ a semiordering of } F\}$$ [10].

By definition, $X_F \neq \emptyset$ iff F is formally real ((1.5) and (1.8)), iff $Y_F \neq \emptyset$ (Corollary (1.15)). Trivially $X_F \subset Y_F$. Note that Theorem (2.12) implies

(6.1) $$X_F = Y_F \text{ iff WH holds in } F .$$

---

9) $(\dot{3})$ follows from $(\dot{1})$.

10) Sometimes we will omit the subscript F.

There is a natural way to introduce a topology on Y_F. For any $a \in \dot{F}$ let

$$H(a) := \{P \in Y_F \mid a \in P\} \quad .$$

Note that $H(1) = Y$, $H(-1) = \emptyset$ and $H(-a) = Y \setminus H(a)$. We consider the topology on Y with subbase $\mathcal{H} := \{H(a) \mid a \in \dot{F}\}$. An open set of Y is generated by elements of the subbase, i.e. is the union of finite intersections of sets $H(a)$. Obviously the generators $H(a)$ are open and closed (clopen). On X_F we consider the induced topology. This topology has subbase $\mathcal{H}_X := \{H_X(a) \mid a \in \dot{F}\}$ where $H_X(a) := H(a) \cap X$.

Later in §10 we will see that there is a one-to-one correspondence between orderings of F and minimal prime ideals of the Wittring $W(F)$. The topology introduced here corresponds to the Zariski topology on $W(F)$. [11)]

Let us first look into the structure of the set $\mathcal{H}_X$. In general $\mathcal{H}_X$ is not closed under intersections. This will turn out to be a very strong condition. But $\mathcal{H}_X$ is closed under symmetric differences. The symmetric difference Δ, defined by

$$A \,\Delta\, B := (A \setminus B) \cup (B \setminus A)$$

for subsets A and B of some given set X, yields a group structure on the power set $\{A \mid A \subset X\}$ of X with $\emptyset$ as zero element.

(6.2) PROPOSITION For all $a,b \in \dot{F}$ $H_X(-ab) = H_X(-a) \,\Delta\, H_X(-b)$. Hence, $\mathcal{H}_X$ is a subgroup of $\{A \mid A \subset X\}$ with respect to Δ.

---

11) This correspondence was introduced independently by Harrison [H] and Leicht-Lorenz [LL].

<u>Proof</u>: If P is an ordering of F, then

$$-ab \in P \iff (-a \in P,\ b \in P) \text{ or } (a \in P,\ -b \in P)$$

for all $a,b \in \dot{F}$.

q.e.d.

By (1.7)(c) $\dot{S}_F$ is a multiplicative subgroup of $\dot{F}$.

(6.3) THEOREM <u>The group</u> $\langle \dot{F}/\dot{S}_F, \cdot \rangle$ <u>is isomorphic to</u> $\langle H_X, \Delta \rangle$. <u>The isomorphism is given by</u> $a\dot{S}_F \mapsto H(-a)$.

<u>Proof</u>: By (6.2) the map $a \mapsto H(-a)$ is a homomorphism from $\dot{F}$ onto H_X. $H(-a) = \emptyset$ iff $H(a) = X$, i.e. iff a is totally positive. Now (1.9) implies that $\dot{S}_F$ is the kernel of this map.

q.e.d.

(6.4) COROLLARY <u>For all</u> $a,b \in \dot{F}$ (1) <u>to</u> (3) <u>are equivalent</u>:

(1) $H(a) = H(b)$

(2) $H_X(a) = H_X(b)$

(3) $a \equiv b \mod \dot{S}_F$.

<u>Proof</u>: (1) ⇒ (2) and (3) ⇒ (1) are trivial. (2) ⇒ (3) follows from Theorem (6.3).

q.e.d.

Next let us look at the topological structure of X and Y resp.

(6.5) THEOREM <u>The space</u> Y_F <u>of all semiorderings of</u> F <u>is totally disconnected and compact</u>. <u>The space</u> X_F <u>of orderings is a closed subspace of</u> Y_F.

<u>Proof</u>: Y is totally disconnected, since two different points $P_1, P_2 \in Y$ can be separated by a clopen set. If $P_1 \neq P_2$, there is an $a \in F$ such that $a \in P_1$ and $-a \in P_2$. Hence, $P_1 \in H(a)$ and $P_2 \in H(-a)$. X is a closed subspace of Y, since every $P \notin X$ is a proper semiordering. Thus there are $a,b \in P$ such that $-ab \in P$.

But then $P \in H(a) \cap H(b) \cap H(-ab)$, which has an empty intersection with X .

It remains to show that Y is compact. To any $P \in Y$ there corresponds uniquely the map

$$\mathrm{sgn}_P : \dot{F} \to \{-1,1\}$$

where $\mathrm{sgn}_P(a) = 1$ iff $a \in P$. Hence, Y corresponds to a certain subset of $\mathcal{I} := \{-1,1\}^{\dot{F}}$ = set of all maps from $\dot{F}$ to $\{-1,1\}$. We endow $\{-1,1\}$ with the discrete topology. Then by the theorem of Tychonoff $\mathcal{I}$ endowed with the product topology is compact. A subbase of this topology is given by the sets

$$H_a^i = \{f \in \mathcal{I} \mid f(a) = i\}$$

where $i \in \{-1,1\}$ and $a \in \dot{F}$. The set Y corresponds to the subset

$$\{f \in \mathcal{I} \mid f^{-1}(\{1\}) \text{ is a semiordering of } F\}$$

of $\mathcal{I}$. We shall identify them. The theorem will be proved after showing that Y is closed in $\mathcal{I}$, since the topology induced by that of $\mathcal{I}$ coincides on Y with the one defined originally. This follows from $H_a^1 \cap Y = H(a)$ and $H_a^{-1} \cap Y = H(-a)$ for all $a \in \dot{F}$.

Now let $f \in \mathcal{I} \setminus Y$. Then $A := f^{-1}(\{1\})$ is not a semiordering of F , i.e. one of the conditions $(\dot{1})$, $(\dot{2})$, $(\dot{4})$ above is violated:

$(\dot{1})$ If $A + A \not\subset A$, there are $a,b \in A$ such that $a+b \notin A$. This implies that f belongs to $H_a^1 \cap H_b^1 \cap H_{a+b}^{-1}$.

$(\dot{2})$ If $\dot{F}^2 \cdot A \not\subset A$, there are $a \in A$ and $b \in \dot{F}$ such that $ab^2 \notin A$. Then f belongs to $H_a^1 \cap H_{ab^2}^{-1}$. If $1 \notin A$, f belongs to H_1^{-1} .

$(\dot{4})$ If $A \cup -A \neq \dot{F}$, there is $a \in \dot{F}$ such that $a \notin A \cup -A$. Then f belongs to $H_a^{-1} \cap H_{-a}^{-1}$.

In all cases we obtain an open neighbourhood of f , disjoint from Y .

q.e.d.

By this theorem the spaces X_F and Y_F are boolean spaces, i.e. totally disconnected and compact topological spaces. It is now natural to ask: what boolean spaces do occur? The answer, given by T.C. Craven in [Cr] is: every one! We will present next his result using, however, another construction. Essential for Craven's proof (and also the one presented here) are the so-called SAP-fields. These fields were introduced by Knebusch, Rosenberg and Ware in [KRW] . They will be discussed and characterized in § 9. A formally real field F is called an SAP-field (SAP = Strong Approximation Property) if every two disjoint closed subsets of X_F can be separated by some $H(a)$. If A and B are subsets of Y_F we shall say that $H(a)$ or simply a separates A from B if $A \subset H(a)$ and $B \subset H(-a)$, i.e. $a \in P$ for all $P \in A$ and $-a \in P$ for all $P \in B$. The proof of the following proposition is an easy exercise on boolean spaces.

(6.6) PROPOSITION *Let* Z *be a boolean space with a subbase* $\mathcal{B}$ *closed under complements. Then* (1)-(3) *are equivalent:*

(1) *Every two disjoint closed subsets of* Z *can be separated by some* $B \in \mathcal{B}$,

(2) $\mathcal{B}$ *is closed under finite intersections,*

(3) *every clopen subset of* Z *belongs to* $\mathcal{B}$.

Furthermore (1') *and* (2') *are equivalent:*

(1') *Every point can be separated from a closed subset of* Z , *not containing this point, by some* $B \in \mathcal{B}$,

(2') $\mathcal{B}$ *is a base of* Z .

Trivially (1) $\Rightarrow$ (1') . The converse is not true in general.

(6.7) THEOREM *Let* F *be an SAP-field and* A *a closed subset of* X_F . *Then there is an algebraic extension* F_1 *of* F *such that the restriction map* $P \mapsto P \cap F$ *is a homeomorphism of* X_{F_1} *and* A .

Proof: Without loss of generality we may assume that $A \neq X, \emptyset$.

Since F is an SAP-field, the sets $H_X(a)$ form a base of topology by (6.6). Hence $X \setminus A = \bigcup_{c \in C} H(c)$ for some $C \subset \dot{F}$ with $H(c) \neq \emptyset$. Fix some $P_0 \in A$ and let

$$F_1 := F(\{\sqrt[2^n]{-c} \mid c \in C,\ n \in \mathbb{N}\})$$

where the roots are taken in a fixed real algebraic closure of $\langle F,P_0 \rangle$. This is possible since $P_0 \notin H(c)$, hence $-c \in P_0$. No $P \in \bigcup_{c \in C} H(c)$ extends to F_1 since some $c \in C$ belongs to P. The restriction map $P \mapsto P \cap F$ from X_{F_1} to X_F is injective. To show this, let $F_2 \subset F_1$ be a maximal extension of F such that the restriction map from X_{F_2} to X_F is injective (there exists one by Zorn's Lemma). If $F_2 \neq F_1$, then there is $c \in C$ such that $F_3 := F_2(\{\sqrt[2^n]{-c} \mid n \in \mathbb{N}\})$ is a proper extension of F_2 in F_1. We claim that the restriction map from X_{F_3} to X_F is injective. Suppose not. Then there are two different orderings P_1 and P_2 of F_3 and a first $n \in \mathbb{N}$ such that $P_1 \cap F_2(\sqrt[2^n]{-c}) = P_2 \cap F_2(\sqrt[2^n]{-c}) =: P'$ and $P'' := P_1 \cap F_2(\sqrt[2^{n+1}]{-c}) \neq P_2 \cap F_2(\sqrt[2^{n+1}]{-c}) =: P'''$. Since the ordering P' of $F_2(\sqrt[2^n]{-c})$ has exactly two extensions to $F_2(\sqrt[2^{n+1}]{-c})$, P'' and P''' constitute these. Let $\sqrt[2^{n+1}]{-c} \in P''$ and $-\sqrt[2^{n+1}]{-c} \in P'''$. Then, going one step further to $F_2(\sqrt[2^{n+2}]{-c})$, P''' obviously cannot extend (contradiction!). Hence $F_2 = F_1$. Thus the map $P \mapsto P \cap F$ is a bijection from X_{F_1} to A, since every $P' \in A$ obviously extends to F_1 ($-c \in P'$ for all $c \in C$).

It remains to show that this map is continuous. Then it is also open since X_{F_1} is compact. The continuity follows directly from

$$H_{X_F}(a) \cap A = \{P \cap F \mid P \in H_{X_{F_1}}(a)\}$$

for all $a \in F$.

q.e.d.

(6.8) LEMMA Let R be real closed. Then there is an algebraic extension F of $R(x)$ such that X_F is homeomorphic to $\{1,-1\}^R$, where $\{1,-1\}^R$ is endowed with the product topology induced by the discrete topology on $\{1,-1\}$.

Proof: Let P_0 be the ordering of the rational function field $R(x)$ where the highest coefficient of a polynomial $f \in R[x]$ is positive in R iff $f \in P_0$. Obviously such an ordering exists. Let

$$F := R(x)(\{\sqrt{x-a} \mid a \in R\})$$

where the root is taken in a fixed real algebraic closure of $\langle R(x),P_0\rangle$. Any ordering P of F is an extension of P_0, since every polynomial of $R[x]$ splits into factors of the type $(x-a)^2 + b^2$ and $x-a$. Since $x-a$ is a square in F, every monic polynomial of $R[x]$ belongs to P. This implies $P \cap R(x) = P_0$. We define a map $\tau: X_F \to \{1,-1\}^R$ by sending $P \in X_F$ to $\sigma(P): R \to \{1,-1\}$ determined by $\sigma(P)(a) = 1 \Longleftrightarrow \sqrt{x-a} \in P$ for all $a \in R$.

Note that $R(x)(\sqrt{x-a_1},\ldots,\sqrt{x-a_n})$ has exactly 2^n order-extensions of P_0 (with $a_1,\ldots,a_n$ distinct). These 2^n orderings are determined by the signs of $\sqrt{x-a_i}$ $(1 \leq i \leq n)$. This can be easily proved by induction, using the fact that $\sqrt{x-a} \notin R(x)(\sqrt{x-a_1},\ldots,\sqrt{x-a_m})$ if $a \notin \{a_1,\ldots,a_m\}$.

From this remark the injectivity of τ follows at once. To prove the surjectivity, fix $\sigma: R \to \{1,-1\}$ and consider the field $F' := R(x)(\{\sqrt{\sigma(a)\sqrt{x-a}} \mid a \in R\})$. If we can extend P_0 to an ordering P' of F', then $P = P' \cap F$ is an ordering of F, where $\sigma(a)\sqrt{x-a}$ is positive. This then proves the surjectivity of τ. If P_0 would not extend to F', then it also would not extend to some subfield

$$F'' := R(x)(\sqrt{\sigma(a_1)\sqrt{x-a_1}},\ldots,\sqrt{\sigma(a_n)\sqrt{x-a_n}})$$

(using the criterion (1.24)). But by the above remark there is an extension P of P_0 to $R(x)(\sqrt{x-a_1},\ldots,\sqrt{x-a_n})$ such that $\sigma(a_i)\sqrt{x-a_i} \in P$ for all $1 \le i \le n$. Iterating (1.26) we thus obtain an ordering of F'' extending P_0 .

It remains to show that τ is continuous. Then τ is a homeomorphism since X_F is compact. The continuity follows from
$\tau^{-1}(\{\sigma:R \to \{1,-1\} \mid \sigma(a) = k\}) = \{P \in X_F \mid k\sqrt{x-a} \in P\} = H_{X_F}(k\sqrt{x-a})$
where $k = 1$ or -1 .

q.e.d.

Now we can prove Craven's theorem using a result which will be proved later (compare (9.4)).

(6.9) THEOREM (Craven) For any boolean space Z there is a field F such that Z is homeomorphic to X_F .

Proof: We notice first the well-known fact that Z is homeomorphic to a closed subset of some space $\{1,-1\}^R$ endowed with the product topology induced by the discrete topology on $\{1,-1\}$. R is a set of suitable infinite cardinality. But there are real closed fields R of any cardinality. Simply take $\mathbb{Q}$, adjoin suitably many indeterminates and take some real algebraic closure. By (6.8), $\{1,-1\}^R$ is homeomorphic to X_F where F is an algebraic extension of $R(x)$. Hence Z is homeomorphic to some closed subset A of X_F . As will be shown in (9.4), every algebraic extension of $R(x)$ is an SAP-field. Then by (6.7), Z is homeomorphic to X_{F_1} where F_1 is an algebraic extension of F .

q.e.d.

From this result we may conclude that the topological structure of X_F does not give too much insight into the set of all orderings of a field F . A much stronger method is described in the following paragraph.

§ 7. REAL PLACES

Semiorderings at a Real Place

From now on we shall deal with (non-archimedean) valuations of a given field F. The reader not familiar with valuation theory may consult [E] or [Ri].

Let G denote an abelian group (written additively). We call $\leq$ an <u>ordering of</u> G if $\leq$ linearly orders G and

$$a \leq b \Rightarrow a+c \leq b+c .$$

A map $v: \dot{F} \to G$ from $\dot{F}$ <u>onto</u> an ordered abelian group is called a <u>valuation of</u> F if for all $a,b \in \dot{F}$

(1) $v(a \cdot b) = v(a) + v(b)$

(2) $v(a+b) \geq \min\{v(a),v(b)\}$, if $a+b \neq 0$.

The map v is usually extended to 0 by taking $v(0) = \infty$, where ∞ is larger than any element of G and $g + \infty = \infty + g = \infty$. Now (1) and (2) are true for all $a,b \in F$. Some easy consequences are

$$v(1) = 0, \quad v(-a) = v(a)$$
$$v(a) \neq v(b) \Rightarrow v(a+b) = \min\{v(a),v(b)\} .$$

As usual we introduce

$$A_v := \{a \in F \mid v(a) \geq 0\}$$
$$M_v := \{a \in F \mid v(a) > 0\}$$
$$U_v := \{a \in F \mid v(a) = 0\} .$$

A_v is a <u>valuation ring of</u> F, i.e. a subring A containing 1 and satisfying

$$a \notin A \Rightarrow a^{-1} \in A$$

for all $a \in \dot{F}$. M_v is a maximal ideal of A_v, the only one. U_v consists of all units. In particular $F_v := A_v/M_v$ is a field, called the <u>residue class field of</u> v.

For every valuation ring A of F there is a valuation $v: F \to G$ such that $A_v = A$. We say that v corresponds to A. If two valuations $v_1: \dot{F} \to G_1$ and $v_2: \dot{F} \to G_2$ correspond to A, then v_1 and v_2 are <u>equivalent</u>, i.e. there is a group isomorphism $\varphi: G_1 \to G_2$ of G_1 and G_2 such that $v_2(a) = \varphi(v_1(a))$ for all $a \in \dot{F}$. If v is a valuation of F, the ordered pair $\langle F,v\rangle$ will be called a <u>valued field</u>.

We will frequently use the following well-known fact.

(7.1) PROPOSITION - <u>Let</u> $v: F_0(x) \to G \cup \{\infty\}$ <u>be a valuation of the field</u> $F = F_0(x)$ <u>of rational functions over</u> F_0 <u>such that</u> $v(a) = 0$ <u>for all</u> $a \in F_0$. <u>Then</u> $G \cong \mathbb{Z}$ <u>and</u> (replacing G by $\mathbb{Z}$)

(1) <u>there is an irreducible polynomial</u> $f \in F_0[x]$ <u>such that</u> $v(f) = 1$ <u>and</u> $v(g) = 0$, <u>for all</u> $g \in F_0[x]$ <u>relatively prime to</u> f, <u>and</u> $F_v \cong F_0[x]/(f)$ <u>induced by mapping</u> $x + M_v$ <u>to</u> $x + (f)$, <u>or</u>

(2) $v(x^n + a_{n-1}x^{n-1} + \ldots + a_0) = -n$ <u>for all</u> $n \in \mathbb{N}$ <u>and</u> $a_0,\ldots,a_{n-1} \in F_0$ <u>and</u> $F_v \cong F_0$.

As an example let us consider $\mathbb{R}(x)$. The irreducible polynomials of $\mathbb{R}[x]$ are of the type $x-a$ for some $a \in \mathbb{R}$ or $(x-a)^2 + b^2$ for some $a,b \in \mathbb{R}$, $b \neq 0$ (compare Lemma (3.4)). In case $v: \mathbb{R}(x) \to \mathbb{Z} \cup \{\infty\}$ corresponds to $x-a$, the residue class field is $\mathbb{R}$. If v corresponds to $(x-a)^2 + b^2$, it is $\mathbb{C}$. Finally, if v is given by (7.1)(2), i.e. $v(\frac{1}{x}) = 1$, then the residue class field is again $\mathbb{R}$. The valuations corresponding to $x-a$ $(a \in \mathbb{R})$ and $\frac{1}{x}$ are called the real places of $\mathbb{R}(x)$. Indeed they correspond to the "real places" a $(a \in \mathbb{R})$ and ∞ on the Riemann Sphere $\mathbb{C} \cup \{\infty\}$.

In general, $v: \dot{F} \to G$ is called a <u>real place of</u> F if the residue class field F_v is formally real and (to avoid trivial cases) $G \neq \{0\}$. The last condition is equivalent to $A_v \neq F$.

A semiordering $\leq$ of F will be called compatible with a valuation v of F (or v compatible with $\leq$) if

$$0 < a \leq b \Rightarrow v(a) \geq v(b)$$

for all $a,b \in F$. Mostly we use a contraposition of this condition:

$$0 < a, \quad v(a) < v(b) \Rightarrow b < a .$$

A symmetric subset A of F (i.e. $a \in A \Rightarrow -a \in A$) is called convex with respect to $\leq$ if

$$0 \leq a \leq b \in A \Rightarrow a \in A$$

for all $a,b \in F$.

(7.2) LEMMA - Let v be a valuation of F and $\leq$ an ordering of F. Then (1) to (4) are equivalent:

(1) $\leq$ is compatible with v

(2) A_v is convex with respect to $\leq$

(3) M_v is convex with respect to $\leq$

(4) $0 \leq a \in M_v \Rightarrow a < 1$.

Proof: (1) $\Rightarrow$ (2): If $0 \leq a \leq b \in A_v$, then (1) implies $v(a) \geq v(b) \geq 0$. Hence $a \in A_v$.

(2) $\Rightarrow$ (3): Let $0 < a < b \in M_v$. Then $b^{-1} < a^{-1}$ and $b^{-1} \notin A_v$ implies $a^{-1} \notin A_v$. But then $a \in M_v$.

(3) $\Rightarrow$ (4): Is trivial, since $1 \leq a$ would imply $1 \in M_v$.

(4) $\Rightarrow$ (1): Let $0 < a$ and $v(a) < v(b)$. Then $0 < v(b\,a^{-1})$, i.e. $b\,a^{-1} \in M_v$. Now (4) implies $b\,a^{-1} < 1$. Then $b < a$, since $\leq$ is an ordering.

q.e.d

Remark: For all implications except (4) $\Rightarrow$ (1) we only used that $\leq$ is a semiordering.

For every valuation v of F we introduce the sets

$$X_F^v := \{P \in X_F \mid P \text{ compatible with } v\}$$

$$Y_F^v := \{P \in Y_F \mid P \text{ compatible with } v\} .$$

(7.3) PROPOSITION - Let v be a valuation of F. Then Y_F^v and X_F^v are closed subsets of Y_F and X_F respectively.

Proof: If $P \notin Y^v$, then there are $a,b \in \dot{F}$ such that $a, b-a \in P$ and $v(a) < v(b)$. But then $H(a) \cap H(b-a)$ contains P and is disjoint to Y^v.

q.e.d.

To get a full characterization of Y_F^v and X_F^v we introduce the notion of a semisection of v which can be found implicitly in Krull's paper [Kr]. A semisection of a valuation $v:\dot{F} \to G$ is a map $s:G \to \dot{F}$ satisfying

(1) $s(0) = 1$

(2) $v(s(g)) = g$

(3) $s(g_1+g_2) \equiv s(g_1)\cdot s(g_2) \bmod \dot{F}^2$

for all $g, g_1, g_2 \in G$. Trivial consequences are $s(2g) \in \dot{F}^2$ and $s(g_1) \equiv s(g_2) \bmod \dot{F}^2$, if $g_1 \equiv g_2 \bmod 2G$. The map s is called a section of v if equality holds in (3). If for instance $G \cong \mathbb{Z}$, there always exists a section. Let $p \in \dot{F}$ such that $v(p) = 1$. Then $s(n) = p^n$ defines a section. In general there does not always exist a section of v. But there are always semisections of v.

(7.4) PROPOSITION - For every valuation $v: \dot{F} \to G$ there exists a semisection of v.

Proof: We define $s: G \to \dot{F}$ in two steps. Let $s(2g) = a^2$ where a is an arbitrary element of $\dot{F}$ such that $v(a) = g$, in particular let $s(0) = 1$. Now consider the $\mathbb{Z}/2\mathbb{Z}$-vectorspace $G/2G$. Let B be a subset of G such that $\{g + 2G \mid g \in B\}$ yields a base of $G/2G$. To every $g \in B$ choose some $a \in \dot{F}$ such that $v(a) = g$.

For every $g \in G$ there are $g_1, \ldots, g_n \in B$ such that $g = g_1 + \ldots + g_n + 2g'$ for some $g' \in G$. Hence we define $s(g) =: s(g_1) \cdot \ldots \cdot s(g_n) \cdot s(2g')$ one easily checks (1), (2) and (3).

q.e.d.

(7.5) LEMMA - Let $v: \dot{F} \to G$ be a valuation of F with residue class field F_v and $s: G \to \dot{F}$ a semisection of v. Then every semiordering $P \in Y_F^v$ induces a pair of functions

$$\mathcal{T}_P: G/2G \to Y_{F_v} \quad \text{and} \quad \sigma_P: G/2G \to \{1,-1\}$$

defined by $\sigma_P(\bar{g})s(g) \in P$ for all $g \in G$ and

$$b + M_v \in \mathcal{T}_P(\bar{g}) \iff b\, s(g)\, \sigma_P(\bar{g}) \in P$$

for all $b \in U_v$. In particular $\sigma_P(\bar{0}) = 1$.

Proof: Since $g + 2G = g' + 2G$ implies $s(g) \equiv s(g') \bmod \dot{F}^2$, the definition of σ_P is correct. Now let $b', b'' \in U_v$ and $b'' = b' + m$ for some $m \in M_v$. Assume $b''\, s(g)\, \sigma_P(\bar{g}) \in P$. Since $v(b''\, s(g)\, \sigma(\bar{g})) = v(b'') + g = g < v(m) + g = v(m\, s(g)\, \sigma_P(\bar{g}))$ and P is compatible with v, we obtain $b'\, s(g)\, \sigma_P(\bar{g}) = (b'' - m)\, s(g)\, \sigma_P(\bar{g}) \in P$. Thus also the second definition does not depend on the choice of the representative. One easily checks that $\mathcal{T}_P(\bar{g})$ is a semiordering of F_v.

q.e.d.

(7.6) COROLLARY - Let v be a valuation of F. Then $Y_F^v \neq \phi$ implies that $A_v = F$ or v is a real place of F.

Proof: Let $P \in Y_F^v$. Then by (7.5), $\mathcal{T}_P(\bar{0})$ is a semiordering of F_v. Hence F_v is formally real.

q.e.d.

(7.7) LEMMA - Let $v: \dot{F} \to G$ be a real place of F and S be a semisection of v. Then each pair of functions

$\mathcal{P}: G/2G \to Y_{F_v}$ <u>and</u> $\sigma: G/2G \to \{1,-1\}$

<u>such that</u> $\sigma(\bar{0}) = 1$ <u>induces a semiordering</u> $\mathcal{P}^\sigma \in Y_F^v$ <u>defined by</u>

$$a \in \mathcal{P}^\sigma \iff \frac{a}{s(v(a))}\,\sigma(\overline{v(a)}) + M_v \in \mathcal{P}(\overline{v(a)})$$

<u>for all</u> $a \in \dot{F}$.

<u>Proof</u>: $\mathcal{P}^\sigma$ is a semiordering of F compatible with v:

($\dot{1}$) Let $a,b \in \mathcal{P}^\sigma$. Then

$$(\alpha) \qquad \frac{a}{s(v(a))}\,\sigma(\overline{v(a)}) + M_v \in \mathcal{P}(\overline{v(a)}) \quad \text{and}$$

$$(\beta) \qquad \frac{b}{s(v(b))}\,\sigma(\overline{v(b)}) + M_v \in \mathcal{P}(\overline{v(b)})$$

<u>Case 1</u>: $v(a) = v(b)$. Then obviously (α) and (β) imply

$$\frac{a+b}{s(v(a))}\,\sigma(\overline{v(a)}) + M_v \in \mathcal{P}(\overline{v(a)})$$

Thus $\frac{a+b}{s(v(a))} \in U_v$. Hence $v(a+b) = v(s(v(a))) = v(a)$. Replacing $v(a)$ by $v(a+b)$, we obtain $a + b \in \mathcal{P}^\sigma$.

<u>Case 2</u>: $v(a) < v(b)$. Then $v(a \pm b) = v(a)$ and $\pm \frac{b}{s(v(a))} \in M_v$. Now (α) implies $a \pm b \in \mathcal{P}^\sigma$.

<u>Case 3</u>: $v(b) < v(a)$. Now (β) implies $a + b \in \mathcal{P}^\sigma$.

Note that in case 2 we also proved

$$a \in \mathcal{P}^\sigma, \quad v(a) < v(b) \Rightarrow a - b \in \mathcal{P}^\sigma$$

Hence $\mathcal{P}^\sigma$ is compatible with v.

($\dot{2}$) Let $a \in \mathcal{P}^\sigma$ and $b \in \dot{F}$. <u>Multiplying</u> (α) by the square $\frac{b^2\, s(v(a))}{s(v(ab^2))} + M_v$ and using $\overline{v(ab^2)} = \overline{v(a)}$ we obtain $ab^2 \in \mathcal{P}^\sigma$. Clearly $1 \in \mathcal{P}^\sigma$ follows from $s(o) = 1$ and $\sigma(\bar{0}) = 1$.

($\dot{4}$) Follows from the corresponding property of $\mathcal{P}(\bar{g})$.

(7.8) THEOREM - <u>Let</u> $v: \dot{F} \to G$ <u>be a real place of</u> F <u>and</u> $s: G \to \dot{F}$ <u>be a semisection of</u> v. <u>Then the constructions of</u> (7.5) <u>and</u> (7.7)

<u>yield an invertible correspondence between</u> Y_F^V <u>and</u> $\{\mathcal{P} \mid \mathcal{P}: G/2G \to Y_{F_v}\} \times$
$\times \{\sigma \mid \sigma: G/2G \to \{1,-1\},\ \sigma(\bar{0}) = 1\}$ i.e. $(\mathcal{P}_P)^{\sigma_P} = P$ <u>and</u> $(\mathcal{P}_{\mathcal{P}^\sigma}, \sigma_{\mathcal{P}^\sigma}) =$
$= (\mathcal{P}, \sigma)$.

<u>Proof:</u> (I) Let $P \in Y_F^V$ and $\mathcal{P}_P$, σ_P be defined as in (7.5). Then for all $a \in \dot{F}$

$$a \in (\mathcal{P}_P)^{\sigma} \iff \frac{a}{s(v(a))} \sigma_P(\overline{v(a)}) + M_v \in \mathcal{P}_P(\overline{v(a)})$$

$$\iff a\ \sigma_P(\overline{v(a)})^2 = a \in P.$$

Therefore $(\mathcal{P}_P)^{\sigma_P} = P$.

(II) Let $\mathcal{P}: G/2G \to Y_{F_v}$ and $\sigma: G/2G \to \{1,-1\}$ such that $\sigma(\bar{0}) = 1$. Let $\mathcal{P}^\sigma$ be defined as in (7.7). Then for all $g \in G$ the definition of (7.7) yields $\sigma(\bar{g})\ s(g) \in \mathcal{P}^\sigma$. Hence $\sigma_{\mathcal{P}^\sigma} = \sigma$. Furthermore $\mathcal{P}_{\mathcal{P}^\sigma} = \mathcal{P}$, since for all $g \in G$ and $b \in U_v$ we obtain

$$b + M_v \in \mathcal{P}_{\mathcal{P}^\sigma}(\bar{g}) \iff b\ s(g)\ \sigma_{\mathcal{P}^\sigma}(\bar{g}) \in \mathcal{P}^\sigma$$

$$\iff b\ s(g)\ \sigma(\bar{g}) \in \mathcal{P}^\sigma$$

$$\iff \frac{b\ s(g)}{s(g)} \sigma(\bar{g})^2 + M_v \in \mathcal{P}(\bar{g})$$

$$\iff b + M_v \in \mathcal{P}(\bar{g})$$

q.e.d.

(7.9) THEOREM - <u>With the notations of</u> (7.8), $P \in X_F^V$ <u>iff the map</u> $\mathcal{P}_P$ <u>is constant,</u> $\mathcal{P}_P: G/2G \to X_{F_v}$ <u>and</u> σ_P <u>is a character of the group</u> $G/2G$.

<u>Proof:</u> First let $P \in X_F^V$. Then for all $g_1, g_2 \in G$:

$$\sigma_P(\bar{g}_1 + \bar{g}_2) = 1 \iff s(g_1 + g_2) \in P$$

$$\iff s(g_1)\ s(g_2) \in P$$

$$\iff s(g_1), s(g_2) \in P \text{ or } s(g_1), s(g_2) \notin P$$

$$\iff \sigma_P(\bar{g}_1)\sigma_P(\bar{g}_2) = 1\ .$$

Hence σ_P is a character of $G/2G$. For all $b \in U_v$ by (7.5)

$$b + M_v \in \mathcal{P}_P(\bar{g}) \iff b\ s(g)\ \sigma_P(\bar{g}) \in P$$

and

(*) $$b + M_v \in \mathcal{P}_P(2G) \iff b \in P$$

Since $s(g)\,\sigma_P(\bar{g}) \in P$ and P is an ordering both conditions are equivalent. Thus $\mathcal{P}_P$ is a constant function. Moreover (*) implies that $\mathcal{P}_P(\bar{0})$ is an ordering of F_v.

Next assume σ_P is a character and $\mathcal{P}_P$ maps $G/2G$ constant into X_{F_v}. Let $a_1, a_2 \in P$. By (7.7) this implies

$$\frac{a_i}{s(v(a_i))}\,\sigma_P(\overline{v(a_i)}) + M_v \in \mathcal{P}_P(\overline{v(a_i)}) = \mathcal{P}_P(\bar{0})$$

Multiplying both we obtain $a_1 \cdot a_2 \in P$, since

$$s(v(a_1a_2)) \equiv s(v(a_1))\,s(v(a_2)) \bmod \dot{F}^2.$$

q.e.d.

<u>Remark</u>: The bijective map $(\mathcal{P},\sigma) \longrightarrow \mathcal{P}^\sigma$ in Theorem (7.8) is also a homeomorphism, if we consider the product topology on $\{\mathcal{P} \mid \mathcal{P}: G/2G \to Y_{F_v}\}$ induced by the topology of Y_{F_v}, and the product topology on $\{\sigma \mid \sigma : G/2G \to \{1,-1\}\}$ induced by the discrete topology on $\{1,-1\}$.

To show this, let $a \in \dot{F}$ and $g := v(a)$, $b := \frac{a}{s(v(a))}$. Then obviously

$$a \in \mathcal{P}^\sigma \iff (\sigma(\bar{g}) = 1 \text{ and } b + M_v \in \mathcal{P}(\bar{g}))$$
$$\text{or } (\sigma(\bar{g}) = -1 \text{ and } -b + M_v \in \mathcal{P}(\bar{g})).$$

This implies that the map $(\mathcal{P},\sigma) \longrightarrow \mathcal{P}^\sigma$ is continuous, hence a homeomorphism.

(7.10) COROLLARY - <u>Let</u> v <u>be a real place of</u> F. <u>Then</u> X_F^v <u>is not empty. Every</u> $P \in Y_F^v$ <u>is non-archimedean</u>.

<u>Proof</u>: Since $v\colon \dot{F} \to G$ is a real place, $X_{F_v} \neq \emptyset$. Choose any constant map $\mathcal{P}\colon G/2G \to X_{F_v}$ and the constant character of $G/2G$. Then (7.8) and (7.9) imply $X_F^v \neq \emptyset$.

Next let $P \in Y_F^v$ and $g \in G$, $g < 0$. Choose $a \in P$ such that $v(a) = g$. Since F_v is formally real, $v(n) = 0$ for all $n \geq 1$. Now

$a \in P$ and $v(a) < v(n)$ imply $a - n \in P$, since P is compatible with v. Thus P is non-archimedean.

(7.11) COROLLARY - Let $v: \dot{F} \to G$ be a real place of F. Then $X_F^v = Y_F^v$ iff $(|G/2G| = 2$ and $|X_{F_v}| = 1)$ or $(|G/2G| = 1$ and $X_{F_v} = Y_{F_v})$.

Proof: Fix a semisection $s: G \to F$. Then using (7.8) and (7.9) the statement of the corollary follows from

a) every map $\sigma: G/2G \to \{1,-1\}$ such that $\sigma(\bar{0}) = 1$ is a character iff $|G/2G| \le 2$

b) if $|X_{F_v}| = 1$, then $|Y_{F_v}| = 1$ (by (1.11) and (1.17)). Therefore every $\mathcal{T}$ is a constant map into X_{F_v}.

q.e.d.

These theorems give a good insight into the set Y_F^v of semiorderings at a real place v of F. Let us now compare different real places. There is one central question: do all sets Y_F^v cover the totality of non-archimedean semiorderings? The same question arises for the space of orderings of F.

The Collection of all Real Places

For a given field F we consider the collection of all real places. Our first theorem will show that every non-archimedean ordering belongs to some X_F^v, where v is a real place of F. We will also give a counterexample in the case of semiorderings.

We need a definition and a lemma. If $\le$ is a semiordering of F and $A \subset B$ are subsets of F, then A is called cofinal in B with respect to $\le$ if for every $b \in B$ there is $a \in A$ such that $b \le a$. Let F_o be a subfield of F then

(7.12) $$A_{F_o}^{\le} := \{a \in F \mid |a| \le b \text{ for some } b \in F_o\}$$

is the smallest convex symmetric subset B of F such that F_o is cofinal in B.

(7.13) LEMMA - Let $\leq$ be a semiordering of F and F_o a subfield of F. Then $A_{F_o}^{\leq}$ is a valuation ring of F, convex with respect to $\leq$.

Proof: Obviously $\mathbb{Q} \subset A_{F_o}^{\leq}$ and $A_{F_o}^{\leq}$ is closed under addition and subtraction. Using the identity $ab = (\frac{a+b}{2})^2 - (\frac{a-b}{2})^2$ it suffices to show that $A_{F_o}^{\leq}$ is closed under squaring to get closedness under multiplication. Let $a \in A_{F_o}^{\leq}$. There is $b \in F_o$ (without loss of generality $1 < b$) such that $|a| \leq b$. Now (1.18) (4) implies $|a| \leq b^2$ and (1.18) (7) implies $|a^2| = |a|^2 \leq b^4 \in F_o$. Hence $a^2 \in A_{F_o}^{\leq}$. Finally $A_{F_o}^{\leq}$ is a valuation ring of F, since $a \notin A_{F_o}^{\leq}$ and (without loss of generality) $1 < a$ imply $0 < a^{-1} < 1$ by (1.18)(3). Thus $a^{-1} \in A_{F_o}^{\leq}$.

q.e.d.

(7.14) THEOREM - The topological space X_F of all orderings of F is the union of the set of all archimedean orderings of F and the sets X_F^v, where v is a real place of F.

Proof: Let $\leq$ be a non-archimedean ordering of F. By (7.13), $A_{\mathbb{Q}}^{\leq}$ is a valuation ring of F, different from F. Let $v: \dot{F} \to G$ correspond to $A_{\mathbb{Q}}^{\leq}$, i.e. $A_v = A_{\mathbb{Q}}^{\leq}$. Since A_v is convex with respect to $\leq$, it is also compatible with $\leq$ (by (7.2)). Hence $\leq$ belongs to X_F^v and v is a real place of F (by (7.6)).

q.e.d.

In the case of a semiordering $\leq$ we cannot conclude that $\leq$ always belongs to some Y_F^v, as we will see later. In general we only get:

(7.15) LEMMA - Let $P \subset F$ be a semiordering of F and F_o a subfield of F, not cofinal in F with respect to P. Let the valuation v <u>of</u> F <u>correspond to</u> $A^P_{F_o}$. <u>Then</u> $(A_v \cap P) / M_v$ <u>is a semiordering of</u> F_v. F_o <u>maps isomorphically to a cofinal subfield of</u> F_v.

<u>Proof</u>: By (7.13) $A^P_{F_o} = A_v$ is convex with respect to P. This implies that M_v is convex with respect to P (see the remark after (7.2)). Hence the quotient $F_v = A_v / M_v$ of A_v by the convex subgroup M_v is ordered as an additive group by $(A_v \cap P) / M_v$. Obviously this is also a semiordering.

q.e.d.

(7.16) COROLLARY - F <u>admits a non-archimedean semiordering iff it admits a non-archimedean ordering</u>.

<u>Proof</u>: Choose $F_o = \mathbb{Q}$ in (7.15) and use (7.10).

q.e.d.

(7.17) THEOREM - <u>On the field</u> $F = \mathbb{Q}(x_1, x_2, \ldots)$ <u>of rational functions over</u> $\mathbb{Q}$ <u>in countably many indeterminates there is a non-archimedean semiordering compatible with no real place of</u> F.

<u>Proof</u>: The idea is to construct a semiordering step by step. In each step we will use a construction by some real place which will be "destroyed" in the next step.

First fix an ordering P' of F such that x_i, $1-x_i \in P'$ for all $i \in \mathbb{N} \setminus \{0\}$, e.g. let P' be the induced ordering from $\mathbb{R}$ if F is mapped to a subfield of $\mathbb{R}$ such that x_i lie between 0 and 1.

For all $n \geq 1$ let $F_n := \mathbb{Q}(x_1, \ldots, x_n)$ and let $F_o := \mathbb{Q}$. We construct a semiordering of F starting with the unique ordering P_o of $F_o = \mathbb{Q}$ and extending the already constructed semiordering P_n of F_n to F_{n+1}. To do this let $v_n : F_n(x_{n+1}) \to \mathbb{Z} \cup \{\infty\}$ be the real place defined in (7.1) (2), where $v_n(x_{n+1}) = -1$ and $v_n(a) = 0$ for all $a \in \dot{F}_n$. The residue class field of v_n is isomorphic to F_n and may be identified with this field. To use the construction of (7.7) let $s: \mathbb{Z} \to F_{n+1}$ be defined by $s(m) = x_{n+1}^{-m}$. This actually is a section of v_n. Let $\sigma: \mathbb{Z} / 2\mathbb{Z} \to \{1,-1\}$ be the trivial character. Let us define the map $\mathcal{P} : \mathbb{Z} / 2\mathbb{Z} \to Y_{F_n}$ by $\mathcal{P}(\bar{0}) := P_n$ and $\mathcal{P}(\bar{1}) := P' \cap F_n$. Let $P_{n+1} := \mathcal{P}^{\sigma}$ be the resulting semiordering of F_{n+1}.

For all $a \in \dot{F}_n$, $v_n(a) = 0$. Hence

$$a \in P_{n+1} \Longleftrightarrow a \in P_n .$$

Thus P_{n+1} extends P_n. Moreover $v_n(x_{n+1} - a) = -1$ implies $x_{n+1} - a \in$ $\in P_{n+1}$, since $\frac{x_{n+1}-a}{x_{n+1}} \equiv 1 \mod M_{v_n}$ implies $\frac{x_{n+1}-a}{x_{n+1}} + M_{v_n} \in (-\bar{1}) = P' \cap F_n$. By the same argument we get $x_{n+1}\, x_n,\ x_{n+1}(1-x_n) \in P_{n+1}$ (recall $1-x_n \in P'$!).

Now let $P = \bigcup_{n \in \mathbb{N}} P_n$, which is obviously a semiordering of $F = \mathbb{Q}(x_1, x_2, \ldots)$. By the construction of P the set $\{x_n \mid n \geq 1\}$ is cofinal in F with respect to P and

$$x_{n+1}\, x_n,\ x_{n+1}(1-x_n) \in P .$$

Now suppose P is compatible with some real place v of F. Then the valuation ring A_v is not cofinal in F with respect to P. Hence there is some $n \geq 1$ such that $x_n \notin A_v$. Now $v(x_n) < v(1) = 0$ implies $v(x_{n+1}\, x_n) < v(x_{n+1})$. But then the compatibility of P with v and $x_{n+1}\, x_n \in P$ would lead to $x_{n+1}\, x_n - x_{n+1} = x_{n+1}(x_n - 1) \in P$. (Contradiction!)

q.e.d.

The next theorem gives some information on the interrelationship of different real places of F.

(7.18) THEOREM - <u>Let</u> P <u>be a semiordering of</u> F, v_1 <u>and</u> v_2 <u>real places of</u> F. <u>Then</u>

(1) <u>all valuation rings of</u> F, <u>convex with respect to</u> P <u>are ordered linearly by inclusion and</u> $A_{\mathbb{Q}}^{P}$ <u>is the smallest one</u>.

(2) $A_{v_1} \subset A_{v_2}$ <u>and</u> A_{v_1} <u>compatible with</u> P <u>imply</u> A_{v_2} <u>compatible with</u> P,

(3) $A_{v_1} \subset A_{v_2}$ <u>implies</u> $Y_F^{v_1} \subset Y_F^{v_2}$,

(4) $Y_F^{v_1} \cap Y_F^{v_2} \neq \emptyset$ <u>implies</u> $Y_F^{v_1} \subset Y_F^{v_2}$ or $Y_F^{v_2} \subset Y_F^{v_1}$.

<u>Proof</u>: (1) is trivial.

(2) Let $a \in P$ and $v_2(a) < v_2(b)$. Then $v_2(ab^{-1}) < 0$ implies $ab^{-1} \notin A_{v_2}$. Thus $ab^{-1} \notin A_{v_1}$, hence $v_1(a) < v_1(b)$. But this implies $a - b \in P$.

(3) follows directly from (2).

(4) follows from (1) and (3).

q.e.d.

We are now going to prove a converse of (7.13), i.e. that every valuation ring of a real place looks like some $A_{F_0}^{\leq}$. First let us recall some more notions and standard results of general valuation theory.[12)]

<u>Remark</u>: Let $v\colon \dot{F} \to G$ be a valuation of F. Then obviously every subfield $F_0 \subset A_v$ maps isomorphically to F_v by the residue class homomorphism. If F_0 is a maximal subfield of A_v, then F_v is an algebraic extension of the image of F_0.

Let $F \subset F_1$ be an extension of fields. A valuation ring A_1 of F_1 is called an <u>extension</u> of a valuation ring A of F if $A = A_1 \cap F$. In this case there are corresponding valuations $v\colon \dot{F} \to G$ and $v_1\colon \dot{F}_1 \to G_1$ such that G is a subgroup of G_1 and v is the restriction of v_1 to F.

(7.19) <u>PROPOSOTION</u> - <u>If $F \subset F_1$ is an algebraic extension and A_1, A_2 are valuation rings of F_1 extending the same valuation ring of F, then $A_1 \subset A_2$ implies $A_1 = A_2$.</u>

---

12) Compare [E] or [Ri] or do it yourself.

(7.20) LEMMA - Let $F \subset F_1$ be an algebraic extension of fields and P a semiordering of F_1. Then F is cofinal in F_1 with respect to P.

Proof: Suppose F is not cofinal in F_1. Then by (7.13) A_F^P is a valuation ring of F_1 containing F, but different from F_1. On the other hand, $F_1 \cap F = A_F^P \cap F = F$ and (7.19) imply $A_F^P = F_1$, a contradiction.

Using this lemma, the earlier Theorem (1.21) follows now directly from (1.20).

(7.21) THEOREM - Let P be a semiordering of F and A a valuation ring of F, convex with respect to P. Then there is a subfield F_0 of A such that $A = A_{F_0}^P$. If F_0 is a maximal subfield of A, then it is algebraically closed in F and the residue class field of A is algebraic over the image of F_0.

Proof: Let F_0 be a maximal subfield of A. (The existence is guaranteed by Zorn's Lemma.) By the above remark the residue class field A/M is algebraic over F_0/M. In particular (by (7.20)) F_0/M is cofinal in A/M with respect to $(A \cap P)/M$. But then F_0 is cofinal in A, which implies $A = A_{F_0}^P$. Moreover, since F_0 is cofinal in all its algebraic extension in F and F_0 is maximal in A, it is algebraically closed in F.

q.e.d.

The next theorem is an improvement of a well-known theorem of Lang (see [L]).

(7.22) THEOREM - Let v be a real place of F and $P \in X_F$. Then v extends to a real place of the real algebraic closure of $\langle F,P \rangle$ iff $P \in X_F^v$. In case v extends, the extension is unique and its residue class field is isomorphic to the real algebraic closure of

$$\langle F_v, (A_v \cap P)/M_v \rangle .$$

Proof: First let P be an ordering of F. By R denote the real algebraic closure of $\langle F,P \rangle$. Assume v_1 is a real place of R. Since R has only one ordering P_1 and $X_R^{v_1} \neq \emptyset$ by (7.10), $P_1 \in X_R^{v_1}$. Hence A_{v_1} is convex with respect to P_1. If v_1 is an extension of v, $A_v = F \cap A_{v_1}$ is convex with respect to P. Thus $P \in X_F^v$. Now we want to show that v_1 is the only real place of R extending v. Let v_2 be a real place of R extending v as well. Then, by the same argument as above, A_{v_2} is convex with respect to P_1. Hence $A_{v_1} \subset A_{v_2}$ or $A_{v_2} \subset A_{v_1}$. By (7.19) both cases imply $A_{v_1} = A_{v_2}$, since R is algebraic over F.

Now let $P \in X_F^v$. By (7.21), $A_v = A_{F_0}^P$ for some maximal subfield F_0 of A_v. Note that F_v is ordered by $(A_v \cap P)/M_v$ and is an algebraic extension of F_0/M_v. Let R denote the real algebraic closure of $\langle F,P \rangle$, and P_1 its unique ordering. Then the valuation ring $A_{F_0}^{P_1}$ obviously extends $A_{F_0}^{P}$. Let $A_{F_o}^{P_1} = A_{v_1}$ for some corresponding valuation v_1 of R. v_1 is a real place of R extending v. Now let F_1 be the algebraic closure of F_0 in R. Then by (3.13), F_1 is real closed. Since F_0 is cofinal in F_1 (use (7.20)), $F_1 \subset A_{v_1}$. Therefore F_1/M_{v_1} is a real closed subfield of R_{v_1}. Since $\langle F_1/M_{v_1}, (F_1 \cap P_1)/M_{v_1} \rangle$ contains (up to isomorphism over $\langle F_0/M_v, (P \cap F_0)/M_v \rangle$) the ordered residue class field $\langle F_v, (A_v \cap P)/M_v \rangle$ and R_{v_1} is algebraic over F_v, we finally obtain that R_{v_1} is the real algebraic closure of $\langle F_v, (A_v \cap P)/M_v \rangle$.

q.e.d.

§ 8. REAL HENSELIAN FIELDS

In § 7 we saw that the space X_F of all orderings of the field F is the union of the set of archimedean orderings of F and the sets X_F^v, where v is any real place of F. We also proved that the analogue for semi-orderings of F is not true in general. As we shall see now, the sets X_F^v (and also the sets Y_F^v) correspond naturally to X_{F_1} where F_1 is a distinguished extension of F, the henselization of F with respect to v. Moreover, it will turn out that the behaviour of a quadratic form over F with respect to weak isotropy is determined by its behaviour at the real places of F and the archimedean orderings of F. In particular, we shall show that the union of all Y_F^v together with the archimedean orderings is dense in Y_F.

A valued field $\langle F,v\rangle$ is called *henselian* if for every monic polynomial $f \in A_v[x]$ and every $a \in A_v$ such that $f(a) \equiv 0 \bmod M_v$ and $f'(a) \not\equiv 0 \bmod M_v$ there is $b \in A_v$ such that $f(b) = 0$ and $b \equiv a \bmod M_v$. This property is mostly called *Hensel's Lemma*. If a henselian field $\langle F,v\rangle$ is in addition formally real, we call it a *real henselian field*. Let $\langle F_1,v_1\rangle$ be an extension of the valued field $\langle F,v\rangle$ (in particular, the value group of v is a subgroup of that of v_1). This extension is called *immediate* if the value groups coincide and the natural embedding of A_v/M_v into A_{v_1}/M_{v_1} is onto.

The facts on henselian fields we use are the following (for proofs see [E] or [Ri], compare also [Ax]).

(8.1) PROPOSITION (1) $\langle F,v\rangle$ *is henselian iff* v *extends uniquely to every algebraic extension of* F.

(2) *Let* char $F_v = 0$ (in particular, let v be a real place). *Then* $\langle F,v\rangle$ *is henselian iff there is no proper immediate algebraic extension of* $\langle F,v\rangle$.

(3) To every valued field $\langle F,v\rangle$ with char $F_v = 0$ there is a unique (up to isomorphisms over $\langle F,v\rangle$ as valued fields) immediate algebraic extension which is henselian.

The extension of (8.1) (3) is called the henselian closure or henselization of $\langle F,v\rangle$.

(8.2) LEMMA Let $\langle F_1,v_1\rangle$ be an immediate extension of $\langle F,v\rangle$. Then the restriction map $P_1 \to P_1 \cap F$ is a homeomorphism of $Y_{F_1}^{v_1}$ and Y_F^v, and similarly of $X_{F_1}^{v_1}$ and X_F^v.

Proof: Let $v:\dot{F} \to G$ and $v_1:\dot{F}_1 \to G_1$. Since $G = G_1$, we can choose a semisection s of v which is also a semisection of v_1. Then (7.8) implies that the restriction map from $Y_{F_1}^{v_1}$ to Y_F^v is bijective. The continuity follows from

$$\{P_1 \in Y_{F_1}^{v_1} \mid P_1 \cap F \in H(a)\} = H_1(a)$$

for all $a \in \dot{F}$, where H is taken in Y_F and H_1 in Y_{F_1}. By (7.3) Y_F^v and $Y_{F_1}^{v_1}$ are closed subsets of Y_F and Y_{F_1} resp. Therefore, they are compact and the restriction map is a homeomorphism.

To get the corresponding result for X, we use (7.9).

q.e.d.

(8.3) THEOREM If $\langle F_1,v_1\rangle$ is henselian, then $Y_{F_1}^{v_1} = Y_{F_1}$ and $X_{F_1}^{v_1} = X_{F_1}$. If $\langle F_1,v_1\rangle$ is the henselian closure of $\langle F,v\rangle$, then the restriction map $P_1 \to P_1 \cap F$ is a homeomorphism of Y_{F_1} and Y_F^v, and similarly of X_{F_1} and X_F^v.

Proof: Let $<$ be a semiordering of F_1. Let $a,b \in F_1$, $0 < a$ and $v_1(a) < v_1(b)$. Then $ba^{-1} \in M_{v_1}$. Hence, $x^2+x+ba^{-1} \equiv x(x+1) \bmod M_{v_1}$. By Hensel's Lemma there are $c,d \in F_1$ such that $x^2+x+ba^{-1} = (x+c)(x+d)$. This implies $c+d = 1$ and $ba^{-1} = cd$. Hence $(c-c^2)a = b$. But then $0 < a$ implies $0 \le a(\frac{1}{2} - c)^2 = a(\frac{1}{4} - c + c^2) = \frac{1}{4}a-b$. Hence $b \le \frac{1}{4}a<a$. Therefore $<$ is compatible with v_1. The rest of the theorem now follows from (8.1)(3) and (8.2).

q.e.d.

(8.4) COROLLARY <u>Let $\langle F,v\rangle$ be a henselian field. Then F is formally real iff F_v is formally real.</u>

<u>Proof</u>: F_v is formally real iff $Y_F^v \neq \emptyset$ (by (7.6) and (7.7)). But by (8.3) $Y_F^v = Y_F$.

q.e.d.

(8.5) COROLLARY <u>Let $\langle F_1,v_1\rangle$ be the henselian closure of $\langle F,v\rangle$. If v is a real place of F, then v_1 is the only real place of F_1 extending v.</u>

<u>Proof</u>: By (8.1)(3) the extension is immediate. Hence v_1 is a real place of F_1 . Let v_2 be a real place of F_1 extending v . Then there is a semiordering $P \in Y_{F_1}^{v_2} \subset Y_{F_1} = Y_{F_1}^{v_1}$. Now (7.18)(1) and (7.19) imply $v_1 = v_2$.

q.e.d.

(8.6) THEOREM <u>Let $v:\dot{F} \to G$ be a real place of the field F . Then F is real closed iff</u>

(a) G <u>is divisible</u>

(b) F_v <u>is real closed</u>

(c) $\langle F,v\rangle$ <u>is henselian.</u>

<u>Proof</u>: First let F be real closed and P its unique ordering. Since $X_F^v \neq \emptyset$ by (7.10), $P \in X_F^v$. Now (b) follows from (7.22). To prove (a), let $g \in G$ and $a \in P$ such that $v(a) = g$. For any $n \in \mathbb{N}$ there is $b \in F$ such that $a = b^n$. Thus $g = nv(b)$. To prove (c) we use (8.1)(2) The only algebraic extension of F is $F(\sqrt{-1})$ (compare (3.3)). If v admits an immediate extension to $F(\sqrt{-1})$, this field would be formally real by (7.10).

Now assume (a), (b) and (c). By (7.10) there is $P \in X_F^v$. By (7.22) v extends to the real algebraic closure of $\langle F,P\rangle$. This extension has to be immediate, since G is divisible and F_v is already real closed. But then (c) and (8.1)(2) imply that F is real closed.

q.e.d.

(8.7) THEOREM[13] *Let* v *be a real place of* F. *Then* $\langle F,v\rangle$ *is henselian iff* $X_{F_1} = X_{F_1}^{v_1}$ *for all finite algebraic extension* $\langle F_1,v_1\rangle$ *of* $\langle F,v\rangle$.

Proof: If $\langle F,v\rangle$ is henselian then every algebraic extension $\langle F_1,v_1\rangle$ is also henselian by (8.1)(1). Hence $X_{F_1}^{v_1} = X_{F_1}$ by (8.3).

Now assume $X_{F_1}^{v_1} = X_{F_1}$ for all finite algebraic extensions $\langle F_1,v_1\rangle$ of $\langle F,v\rangle$. This obviously implies $X_{F_1}^{v_1} = X_{F_1}$ for all algebraic extensions $\langle F_1,v_1\rangle$ of $\langle F,v\rangle$. Let R be some real algebraic closure of F and P its unique ordering. Then $\{P\} = X_R = X_R^{v_1}$ for *all* extensions v_1 of v to R. Thus every extension of v to R is a real place of R, but by (7.22) or by (8.5) there is only one such real place. Since $\langle R,v_1\rangle$ is henselian by (8.6)(c), v_1 extends also uniquely to $R(\sqrt{-1})$ (by (8.1)(1)). In summary, we proved that v extends uniquely to $R(\sqrt{-1})$. By (8.1)(1), $\langle F,v\rangle$ is henselian.

q.e.d.

Let us now restate Corollary (7.11) for real henselian fields, using Theorem (8.3).

---

13) This theorem was first proved by the author's student W. Berger [B].

(8.8) THEOREM Let $v: \dot{F} \to G$ be a real place of F and $\langle F,v\rangle$ be henselian. Then $X_F = Y_F$ iff ($|G/2G| = 2$ and $|X_{F_v}| = 1$) or ($|G/2G| = 1$ and $X_{F_v} = Y_{F_v}$).

Thus we obtained a characterization of henselian fields satisfying the Weak Hasse Principle WH (compare (2.12)).

To get some examples for Theorem (8.8) let us introduce the concept of fields of formal power series (compare also [Fu]). Let G be an ordered abelian group and F some field. Let

$$F((G)) := \{f: G \to F \mid \sup(f) \text{ is well-ordered}\}$$

where $\sup(f) := \{g \in G \mid f(g) \neq 0\}$ is called the support of f. On the set $F((G))$, sum and product of $f, f' \in F((G))$ are defined by

$$f + f' := \sum_{g\in G} (f_g + f'_g)\, X^g$$

$$f \cdot f' := \sum_{g\in G} \Big(\sum_{g_1+g_2=g} f_{g_1} \cdot f'_{g_2}\Big)\, X^g .$$

Here we used the notation $\sum_{g\in G} f_g X^g$ for an element f of $F((G))$ with $f_g := f(g)$. F is embedded by $a \to a X^0$. The set $F((G))$ together with the above defined operations yields a field (see [Fu]). It is even a valued field by the map

$$v(f) = \text{least } g \in G \text{, such that } f(g) \neq 0 .$$

This valuation is called the natural valuation of $F((G))$. The residue class field is isomorphic to F and the value group is obviously G. As an example consider $\mathbb{R}((\mathbb{Z}))$. Since the well-ordered subsets of $\mathbb{Z}$ are exactly the subsets with a least element, the elements of $\mathbb{R}((\mathbb{Z}))$ are often called formal Laurent series with real coefficients. What makes the fields $F((G))$ of formal power series important is the fact that $\langle F((G)),v\rangle$ is a henselian field (see [E] or [Ri]).

Looking at Theorem (8.8) we see that the fields $\mathbb{R}((\mathbb{Z}))$, $\mathbb{Q}((\mathbb{Z}))$ and $\mathbb{Q}((\mathbb{Z}))((G))$ are examples of fields satisfying WH , if G is 2-divisible (i.e. $|G/2G| = 1$). A field not satisfying WH is for example $\mathbb{Q}(\sqrt{2})((\mathbb{Z}))$, since $\mathbb{Q}(\sqrt{2})$ admits two orderings.

Local-Global Principles

In this section we investigate the connection of the behaviour of quadratic forms over F with all real places of F (and the archimedean orderings).

Let F be a formally real field. In (2.12) we proved that every quadratic form, indefinite with respect to all orderings of F is weakly isotropic in F iff $X_F = Y_F$. The condition $X_F = Y_F$ is very strong, (8.8) for example shows how strong it is for henselian fields. Given a quadratic form $\rho = \langle a_1,\dots,a_n\rangle$ over F which is indefinite with respect to all orderings of F , one may ask what to add in order to obtain its weak isotropy. One answer is Theorem (8.12) below which was conjectured by L. Bröcker and proved independently by L.Bröcker (see [Br]) and the author.

Let $v:\dot{F} \to G$ be a valuation of F and $s:G \to \dot{F}$ a semisection of v . To any regular quadratic form $\rho = \langle a_1,\dots,a_n\rangle$ over F we define certain quadratic forms over the residue class field F_v of v . Since we are here only interested in the question of isotropy, we may assume $\rho = \rho^{(1)} \perp\dots\perp \rho^{(m)}$ with $\rho^{(i)} = \langle a_{i1},\dots,a_{in_i}\rangle$ where a_{ij} are all the a from ρ such that v(a) belongs to a fixed residue class of G/2G. We call the forms

$$\rho_v^{(i)} := \langle a_{i1}\, s(v(a_{i1}))^{-1},\dots,\, a_{in_i}\, s(v(a_{in_i}))^{-1}\rangle$$

$(1 \le i \le m)$ the residue class forms of ρ . Note that $a\, s(v(a))^{-1} \in U_v$

(if $a \neq 0$).

(8.9) THEOREM Let v be a valuation of F, char $F_v \neq 2$ and $\rho = \rho^{(1)} \perp \ldots \perp \rho^{(m)}$ a regular quadratic form over F, where $\rho_v^{(1)}, \ldots, \rho_v^{(m)}$ are the residue forms of ρ.

(1) If ρ is isotropic in F, then some $\rho_v^{(i)}$ is isotropic in F_v.

(2) If some $\rho_v^{(i)}$ is isotropic in F_v and $\langle F,v\rangle$ is henselian, then ρ is isotropic in F.

Proof: (1) Let $\rho = \langle a_1,\ldots,a_n\rangle$, $a_i \in \dot F$ and assume that $\sum_{i=1}^{n} a_i b_i^2 = 0$ for some $b_i \in F$, not all zero. Let $v(a_j b_j^2)$ be minimal for some $j \in \{1,\ldots,n\}$, let a_j belong to $\rho^{(1)}$. Multiplying $\Sigma a_i b_i^2$ by $s(v(a_j))^{-1} b_j^{-2}$ and passing to F_v, we find that $\rho_v^{(1)}$ is isotropic in F_j. This follows, since $v(a_j b_j^2) = v(a_i b_i^2)$ implies $v(a_j) \equiv$ $\equiv v(a_i) \bmod 2G$, and therefore $s(v(a_j)) \equiv s(v(a_i)) \bmod \dot F^2$.

(2) Let $\rho^{(1)} = \langle c_1,\ldots,c_r\rangle$ be a subform of ρ such that the residue class form $\rho_v^{(1)} = \langle c_1\, s(v(c_1))^{-1},\ldots,c_r\, s(v(c_r))^{-1}\rangle$ is isotropic in F_v. Let $b_1,\ldots,b_v \in A_v$, not all in M_v, such that $\sum_{i=1}^{r} c_i\, s(v(c_i))^{-1} b_i^2 \in M_v$. Replacing one $b_j \notin M_v$ by an indeterminate x we get a quadratic polynomial of $A_v[x]$ which has a simple zero in F_v. Hence by Hensel's Lemma it also has a zero in F. Hence $\sum_{i=1}^{r} c_i\, s(v(c_i))^{-1} d_i^2 = 0$ for some $d_i \in F$, not all zero. Since $s(v(c_i)) \equiv s(v(c_j)) \bmod \dot F^2$ for all $1 \leq i, j \leq r$, it follows that $\rho^{(1)}$ and hence ρ is isotropic in F.

q.e.d.

Remark: If we do not assume $\langle F,v\rangle$ to be henselian in (2) we only obtain $0 = \min\{v(c_i\, s(v(c_i))^{-1} b_i^2)\} < v(\sum_{i=1}^{r} c_i\, s(v(c_i))^{-1} b_i^2)$. This implies the existence of $f_i \in F$, not all zero, such that

$$\min\{v(a_i f_i^2)\} < v(\sum_{i=1}^{n} a_i f_i^2) .$$

This theorem in particular shows that for henselian fields $\langle F,v\rangle$ the question of isotropy in F reduces to the corresponding question for the residue class field F_v. In view of this fact one tries to reduce the question of isotropy in an arbitrary field F to that of the henselizations of $\langle F,v\rangle$, where v runs through a suitable set of valuations of F. There is a well-known classical example for this process, the "Hasse-Minkowski Principle" (see [MH]). There, instead of the henselizations, the completions with respect to the topology induced by a valuation are taken. But these completions are also henselian fields.

Let us return to the question of weak isotropy. In this case, only the real places turn out to be relevant, and in connection with real places, also semiorderings. We need some sufficient conditions for a semiordering of F to be compatible with a real place v of F.

(8.10) PROPOSITION *Let* $v:\dot{F} \to G$ *be a valuation of* F *and* $\leq$ *a semiordering of* F *such that* A_v *is convex with respect to* $\leq$. *Then for all* $a,b \in F$ *and* $g \in G$ $0 < a$, $v(a) \neq v(b)$, $v(a) \leq 2g \leq v(b)$ *implies* $b < a$.

Proof: Let $0 \leq b$ and $v(c) = g$ for some $c \in F$. Then $v(a) \leq v(c^2) \leq$ $\leq v(b)$ implies $v(ac^{-2}) \leq 0 \leq v(bc^{-2})$. By $v(a) \neq v(b)$, equality does not hold on both sides simultaneously.

Case 1: $v(ac^{-2}) < 0 \leq v(bc^{-2})$.
This implies $bc^{-2} \in A_v$, but $ac^{-2} \notin A_v$. From $0 < a$ and the convexity of A_v with respect to $\leq$ we obtain $bc^{-2} < ac^{-2}$. Hence $b < a$.

Case 2: $v(ac^{-2}) \leq 0 < v(bc^{-2})$.
This implies $bc^{-2} \in M_v$, but $ac^{-2} \notin M_v$. By the remark after (7.2), M_v is also convex with respect to $\leq$. Hence once more we get $bc^{-2} <$ $< ac^2$, and thus $b < a$.

q.e.d.

(8.11) COROLLARY *Let* $v:\dot{F} \to G$ *be a valuation of* F *and* P *a semi-ordering of* F *such that* A_v *is convex with respect to* P. *If for all* $g_1, g_2 \in G$ *with* $g_1 < g_2$ *there is* $g \in G$ *such that* $g_1 \leq 2g \leq g_2$, *then* P *is compatible with* v. *In particular this holds, if* G *is* 2-*divisible or* $G \simeq \mathbb{Z}$.

Proof: follows directly from (8.10).

q.e.d.

The next theorem is a certain *Local-Global Principle* for weak isotropy of regular quadratic forms.

(8.12) THEOREM *Let* F *be formally real and* ρ *a quadratic form over* F. *If* ρ *is indefinite with respect to all orderings of* F *and weakly isotropic in all henselizations with respect to real places* $v:\dot{F} \to G$, *where* G *is not* 2-*divisible, then* ρ *is weakly isotropic in* F.

Proof: Suppose $\rho = \langle a_1, \dots, a_n \rangle$ is not weakly isotropic in F. Then by (2.9) there is a proper semiordering P of F such that ρ is definite with respect to P. Let us assume without loss of generality that $a_i \in P$ for all $1 \leq i \leq n$. By (1.20) P is non-archimedean. Hence to the valuation ring $A_{\mathbb{Q}}^{P}$ there corresponds some real place $v:\dot{F} \to G$ of F (compare (7.15)). Unfortunately P may not be compatible with v. Hence we may not be able to extend P to the henselian closure of $\langle F,v \rangle$ in general. However, we shall use Proposition (8.10).

Let us first assume that all $v(a_i)$ belong to $2G$. Multiplying by suitable squares, we may assume $a_i \in U_v$ for all $1 \leq i \leq n$. By (7.15) and (1.20) $(P \cap A_v)/M_v$ is an (archimedean) ordering of F_v. Using a suitable construction of (7.7) and Theorem (7.9) we obtain an ordering P' of F such that $(P' \cap A_v)/M_v = (P \cap A_v)/M_v$. But then

$a_i \in P'$ for all $1 \le i \le n$. Hence ρ is definite with respect to the ordering P' of F, contradicting the assumption of the theorem.

Now let us assume that not all $v(a_i)$ belong to $2G$. For every $1 \le i \le n$ we let G_i consist of all $g \in G$ such that there are no $n \in \mathbb{N}$ and $g_1 \in G$ with $0 \le v(a_i) + 2g_1 < n|g|$. Obviously G_i is a convex subgroup of G. Thus also $G' := \bigcup_{i=1}^{n} G_i$ is a convex subgroup of G. Hence we can order $G^* := G/G'$ by

$$g_1 + G' < g_2 + G' :\Longleftrightarrow g_1 < g_2 \quad \text{and} \quad g_2 - g_1 \notin G'.$$

Now the definition $v^*(a) = v(a) + G'$ leads to a valuation $v^*: F \to G^*$. This is a real place of F too. Indeed, take some ordering $P' \in X_F^v$, then by (7.18)(2) and (7.2), A_{v^*} is compatible with P'. Hence by (7.6) v^* is a real place, since by construction G^* is not 2-divisible, in particular $G^* \neq \{0\}$. Therefore, by the assumption of the theorem, $m\rho$ is isotropic in the henselian closure of $\langle F, v^* \rangle$ for some $m \ge 1$. Thus by (8.9)(1) some residue class form of $m\rho$ is isotropic in F_{v^*}. Using the remark after (8.9) this in turn implies the existence of some $f_j \in F$, not all zero, such that

$$\min\{v^*(c_j f_j^2)\} < v^*\left(\sum_{j=1}^{mn} c_j f_j^2\right)$$

where $c_j \in \{a_1, \ldots, a_n\}$ for all $1 \le j \le mn$. Let $v^*(c_i f_i^2)$ be minimal. Then

$$v^*(c_i) < v^*\left(\sum_{j=1}^{mn} c_j \left(\frac{f_j}{f_i}\right)^2\right).$$

Hence there is $i \in \{1, \ldots, n\}$ and $p \in P$ such that

$$v^*(a_i) < v^*(a_i + p).$$

This implies $v(a_i) < v(a_i + p)$ and $g_0 := v(a_i + p) - v(a_i) \notin G'$. From this we shall deduce the existence of some $g \in G$ such that

$$(*) \qquad v(a_i) \le 2g \le v(a_i + p).$$

Using the Proposition (8.10) we then obtain $a_i - (a_i+p) = -p \in \dot{P}$, a contradiction.

It remains to prove (*). Since $g_o \notin G'$, there are $m \in \mathbb{N}$ and $g_1' \in G$ such that

$$0 \leq v(a_i) + 2\,g_1' < m\,g_o \ .$$

Let m be minimal with this property. Then

$$(m-1)g_o \leq v(a_i) + 2\,g_1' < m\,g_o \ .$$

Substracting an even multiple of g_o, we obtain two possible cases.

<u>Case 1</u>: $0 \leq v(a_i) + 2\,g_1 < g_o = v(a_i+p) - v(a_i)$.
This implies $v(a_i) \leq 2(v(a_i)+g_1) < v(a_i+p)$.

<u>Case 2</u>: $g_o \leq v(a_i) + 2\,g_1 < 2\,g_o$.
This implies $v(a_i+p) \leq 2(v(a_i)+g_1) < v(a_i+p) + g_o$. Let $g_2 := 2(v(a_i)+g_1) - v(a_i+p)$. Then $v(a_i) < 2(v(a_i)+g_1) - 2\,g_2 \leq v(a_i+p)$, since $g_2 < g_o$.

q.e.d.

From (8.12) we easily deduce a more convenient <u>Local-Global Principle</u>

(8.13) THEOREM <u>Let</u> F <u>be formally real and</u> ρ <u>a quadratic form over</u> F. <u>If</u> ρ <u>is indefinite with respect to all archimedean orderings of</u> F <u>and is weakly isotropic in the henselizations of all real places of</u> F, <u>then</u> ρ <u>is weakly isotropic in</u> F.

<u>Proof</u>: Let $\rho = \langle a_1,\ldots,a_n\rangle$ $(a_i \in \dot{F})$ satisfy the assumption of the theorem. To use Theorem (8.12) we have to show that ρ is indefinite with respect to a given non-archimedean ordering P of F. By (7.14) $P \in X_F^v$ for some real place v of F. By (8.3), P extends to an ordering of the henselization of $\langle F,v\rangle$. Since ρ is weakly isotropic

in this henselization, ρ is indefinite with respect to P.

q.e.d.

Using this Local-Global-Principle we will now prove the following approximation theorem about the space of semiorderings.

(8.14) THEOREM Let F be formally real. Then the archimedean orderings together with the semiorderings compatible with some real place of F form a dense subset of Y_F.

Proof: Let $P' \in Y_F$. Consider some open neighbourhood $H(a_1) \cap \ldots \cap H(a_n)$ of P'. The quadratic form $\rho = \langle 1, a_1, \ldots, a_n \rangle$ is definite with respect to P'. Hence it is not weakly isotropic. But then by (8.13), ρ is definite with respect to some archimedean ordering P (i.e. $1, a_1, \ldots, a_n \in P$), or it is not weakly isotropic in the henselian closure of $\langle F, v \rangle$ for some real place v of F. But then by (2.9) there is a semiordering P'' of this henselian closure such that ρ is definite with respect to P'' (i.e. $1, a_1, \ldots, a_n \in P'' \cap F$). By (8.3), $P = P'' \cap F \in Y_F^v$. In both cases we have $P \in H(a_1) \cap \ldots \cap H(a_n)$.

q.e.d.

§9. SAP-FIELDS

Characterization of SAP-Fields

In this paragraph we will give several characterizations of fields F satisfying $X_F = Y_F$. By Theorem (2.12) $X_F = Y_F$ iff F satisfies the Weak Hasse-Principle:

WH: every quadratic form indefinite with respect to all orderings of F is weakly isotropic in F.

It will turn out in particular that this class of fields coincides with the class of SAP-fields already introduced in §6 (compare (6. 6)) and used to prove that X_F can be any boolean space (Theorem (6.9)). In [KRW] the SAP-fields are characterized by their Wittrings (see (10.16)). In [P] the author gave a characterization of fields satisfying WH in terms of real places. In most cases this characterization can be used to decide whether a given field belongs to this class or not. In [ELP] the equivalence of SAP and WH has been shown by Elman, Lam and the author. In [EL] Elman and Lam proved the equivalence of (1) and (1') in (6.6) for fields F, i.e. in the case $Z = X_F$ and $V = H_X$. Using the characterization of [P], L. Bröcker showed in [BR] that (1') for X_F can be "weakened" equivalently to the case where A contains only 3 elements.

All these results will be combined in the following theorem. The subscript X in (1), ..., (1'), ... of (6.6) indicates that Z is replaced by X and V by H_X, while the subscript Y indicates the replacement of Z by Y and V by H (compare §6).

(9.1) THEOREM - Let F be a formally real field. Then (1_X), $(1'_X)$, (1_Y) and $(1'_Y)$ of (6.6) are equivalent. Moreover they are equivalent to each one of the following conditions:

(a_X) Every two finite disjoint sets of orderings of F can be separated by some $a \in \dot{F}$.

(b_X) Every three non-archimedean orderings of F can be separated by some $a \in \dot{F}$ from any other non-archimedean ordering of F.

(b_Y) Every two non-archimedean orderings of F can be separated by some $a \in \dot{F}$ from any proper semiordering of F.

(c) For all real places $v : \dot{F} \to G$ of F we have $|G/2G| \leq 2$ and, if $|G/2G| = 2$, then $|X_{F_v}| = 1$.

(d) $X_F = Y_F$.

(e) WH holds in F.

Proof : The proof runs as follows:

$$\begin{array}{ccccccccccc} & & & \nearrow (a_X) \searrow & & & & & & & \\ (1_X) & \Rightarrow & (1'_X) & \Rightarrow & (b_X) & \searrow & & & & \nearrow & (1_X) \\ & & & & & & (c) \Rightarrow (d) \Rightarrow (e) & & & & \\ (1_Y) & \Rightarrow & (1'_Y) & \Rightarrow & (b_Y) & \nearrow & & & & \searrow & (1_Y) \end{array}$$

The first six implications are trivial.

(d) $\iff$ (e) has been shown in (2.12).

(b_X) $\Rightarrow$ (c) and (b_Y) $\Rightarrow$ (c) now will be proved together. Suppose (c) does not hold.

Case 1: $|G/2G| \geq 4$ for some real place $v: \dot{F} \to G$. For $g \in G$ let $\bar{g} := g + 2G$. Let $\{\bar{g}_1, \bar{g}_2, \ldots\}$ constitute a base of the $\mathbb{Z}/2\mathbb{Z}$-vector space $G/2G$. Furthermore let $\mathcal{P}: G/2G \to X_{F_v}$ be some constant function, P_o its value and $s: G \to \dot{F}$ some semisection.

First we show the contrary of (b_X). We define four characters $\sigma_i: G/2G \to \{1,-1\}$ by the conditions

	σ_1	σ_2	σ_3	σ_4
$\bar{g}_1$	1	1	-1	-1
$\bar{g}_2$	1	-1	1	-1
others	1	1	1	1

We claim that no $a \in \dot{F}$ separates $\mathcal{P}^{\sigma_1}$ from $\mathcal{P}^{\sigma_2}$, $\mathcal{P}^{\sigma_3}$ and $\mathcal{P}^{\sigma_4}$. Suppose $a \in \dot{F}$ does, i.e. $-a \in \mathcal{P}^{\sigma_1}$ and $a \in \mathcal{P}^{\sigma_i}$ for $2 \leq i \leq 4$.

Let

$$\overline{v(a)} = n_1 \bar{g}_1 + n_2 \bar{g}_2 + \sum_{\bar{g} \neq \bar{g}_1, \bar{g}_2} n_g \bar{g} .$$

Hence $\sigma_i(\overline{v(a)}) = \sigma_i(\bar{g}_i)^{n_1} \cdot \sigma_i(\bar{g}_2)^{n_2}$. By (7.7) we get

$$a \in \mathcal{P}^{\sigma_i} \Leftrightarrow \frac{a}{s(v(a))} \ \sigma_i(\bar{g}_1)^{n_1} \ \sigma_i(\bar{g}_2)^{n_2} + M_v \in P_o$$

Letting $b := \frac{a}{s(v(a))}$ and checking all possibilities for n_1 and n_2 we get

(n_1,n_2)	$-a \in \mathcal{P}^{\sigma_1}$	$a \in \mathcal{P}^{\sigma_2}$	$a \in \mathcal{P}^{\sigma_3}$	$a \in \mathcal{P}^{\sigma_4}$
(0,0)	$b+M_v \notin P_o$	$b+M_v \in P_o$	$\in$	$\in$
(1,0)	$\notin$	$\in$	$\notin$	$\notin$
(0,1)	$\notin$	$\notin$	$\in$	$\notin$
(1,1)	$\notin$	$\notin$	$\notin$	$\in$

In every line there is a contradiction.

Next we show the contrary of (b_Y). In this case σ need not always be a character. Hence we take σ_2, σ_3 as above and define a map $\sigma_1: G/2G \to \{1,-1\}$ by $\sigma_1(\bar{g}_1) = \sigma_1(\bar{g}_2) = \sigma_1(\bar{g}_1+\bar{g}_2) = -1$, $\sigma_1(\bar{0}) = 1$

$$\sigma_1(n_1\bar{g}_1 + n_2\bar{g}_2 + \sum_{\bar{g} \neq \bar{g}_1, \bar{g}_2} n_g \bar{g}) := \sigma_1(n_1\bar{g}_1 + n_2\bar{g}_2) .$$

In case of σ_1 we get

$$a \in \mathcal{P}^{\sigma_1} \Leftrightarrow b \ \sigma_1(n_1\bar{g}_1 + n_2\bar{g}_2) + M_v \in P_o ,$$

if $\overline{v(a)} = n_1\bar{g}_1 + n_2\bar{g}_2 + \ldots$

Now we obtain the following contradictory table

(n_1,n_2)	$-a \in \mathcal{P}^{\sigma_1}$	$a \in \mathcal{P}^{\sigma_2}$	$a \in \mathcal{P}^{\sigma_3}$
(0,0)	$b+M_v \notin P_o$	$\in$	$\in$
(1,0)	$b+M_v \in P_o$	$\in$	$\notin$
(0,1)	$\in$	$\notin$	$\in$
(1,1)	$\in$	$\notin$	$\notin$

By Theorem (7.9) $\mathcal{P}^{\sigma_1}$ is a proper semiordering. By (7.10), $\mathcal{P}^{\sigma_2}$ and

$\mathcal{P}^{\sigma_3}$ are non-archimedean orderings.

<u>Case 2</u>: $|G/2G| = 2$ and P_0, P_1 are different orderings of F_v, where $v:\dot{F} \to G$ is some real place of F. Let $G = 2G \cup (g_1+2G)$ and $s:G \to \dot{F}$ be a semisection of v. Further define characters $\sigma_i:G/2G \to \{1,-1\}$ $(i=1,2)$ by $\sigma_1(\bar{g}_1) = 1$ and $\sigma_2(\bar{g}_1) = -1$. Let $\mathcal{P}_j$ be the constant map to X_{F_v} with value P_j $(j=0,1)$.

<u>First</u> we show the contrary of (b_X). By (7.9), $\mathcal{P}_0^{\sigma_1}, \mathcal{P}_0^{\sigma_2}, \mathcal{P}_1^{\sigma_1}, \mathcal{P}_1^{\sigma_2} \in X_F$. Suppose $-a \in \mathcal{P}_0^{\sigma_1}$ and $a \in \mathcal{P}_0^{\sigma_2}, \mathcal{P}_1^{\sigma_1}, \mathcal{P}_1^{\sigma_2}$. Let $\overline{v(a)} = n\bar{g}_1$ $(n=0,1)$. Then $\sigma_i(\overline{v(a)}) = \sigma_i(\bar{g}_1)^n$. By (7.7) we get

$$a \in \mathcal{P}_j^{\sigma_i} \iff \frac{a}{s(v(a))}\,\sigma_i(\bar{g}_1)^n + M_v \in P_j .$$

If $n = 0$, then $-a \in \mathcal{P}_0^{\sigma_1}$ and $a \in \mathcal{P}_0^{\sigma_2}$ give a contradiction. If $n = 1$, then $a \in \mathcal{P}_1^{\sigma_1}$ and $a \in \mathcal{P}_1^{\sigma_2}$ give a contradiction.

<u>Next</u> we show the contrary of (b_Y). We use $\mathcal{P}_0^{\sigma_1}$ and $\mathcal{P}_1^{\sigma_1}$ from above. Now we can also use the map $\mathcal{P}:G/2G \to X_{F_v}$ defined by $\mathcal{P}(\bar{0}) = P_0$ and $\mathcal{P}(\bar{g}_1) = P_1$. Suppose $-a \in \mathcal{P}^{\sigma_1}$ and $a \in \mathcal{P}_0^{\sigma_1}, \mathcal{P}_1^{\sigma_1}$. Then $-a \in \mathcal{P}^{\sigma_1}$ and $a \in \mathcal{P}_n^{\sigma_1}$ give a contradiction.

(c) $\Rightarrow$ (d): Suppose $P \in Y_F$ is a proper q-ordering. Then there are $a,b \in \dot{F}$ such that $1, a, b, -ab \in P$. Hence $\rho = \langle 1,a,b,-ab\rangle$ is not weakly isotropic in F. Since ρ is indefinite with respect to every ordering of F, by Theorem (8.12) there is a real place $v:\dot{F} \to G$, G not 2-divisible, such that ρ is not weakly isotropic in the henselian closure of $\langle F,v\rangle$. Since G is not 2-divisible, (c) implies $|G/2G| = 2$ and $|X_{F_v}| = 1$. But then by Theorem (8.8) and (2.12), ρ is weakly isotropic in the henselian closure of $\langle F,v\rangle$. (Contradiction!)

(e) $\Rightarrow$ (1_X): Actually we show (e) $\Rightarrow$ (2_X). Let $a,b \in \dot{F}$. We show the existence of some $c \in \dot{F}$ such that $H_X(a) \cap H_X(b) = H_X(c)$. Consider the quadratic form $<1,a,-b,ab>$ which is indefinite with respect to all orderings of F. Hence by (e) it is weakly isotropic, i.e. there are $s_1,s_2,s_3,s_4 \in S_F$, not all zero, such that

$$s_1 + a\, s_2 - b\, s_3 + a\, bs_4 = 0 .$$

Hence $-s_1 + b\, s_3 = a(s_2 + b\, s_4) =: c$. If $c \neq 0$ one easily checks $H_X(a) \cap H_X(b) = H_X(c)$. If $-s_1 + b\, s_3 = 0$ and $s_3 \neq 0$, then $b \in S_F$. In this case $H_X(a) \cap H_X(b) = H_X(a)$. If $-s_1 + b\, s_3 = 0$ and $s_3 = 0$, then $b \in -S_F$ and $H_X(a) \cap H_X(b) = H_X(-1)$.

(e) $\Rightarrow$ (1_Y): We already showed (e) $\Rightarrow$ (1_X). By (2.12) also (e) $\Rightarrow X_F = Y_F$. Hence (e) $\Rightarrow$ (1_Y).

q.e.d.

<u>Remark</u>: Note that (b_X) cannot be "weakened" equivalently to two orderings similar to (b_Y). If $P_1 \neq P_2,P_3$ are orderings of F, then P_1 can be separated from P_2 and P_3. Let a separate P_1 from P_2 and b separate P_1 from P_3, i.e. $a,b \in P_1$ and $-a \in P_2$, $-b \in P_3$. Then a,b or ab separate P_1 from P_2 and P_3.

(9.2) COROLLARY <u>If</u> G <u>is</u> <u>2-divisible for every real place</u> $v:\dot{F} \to G$ <u>of</u> F, <u>then any algebraic extension</u> F_1 <u>of</u> F <u>is an SAP-field. This holds for example if</u> F <u>admits only archimedean orderings or has only one ordering</u>.

<u>Proof</u>: If F has only archimedean orderings, then there is no real place of F (see (7.10)). If F has a unique non-archimedean ordering P and $v:\dot{F} \to G$ is a real place of F, then obviously $\{P\} = X_F^V$. Hence by (7.9), G has to be 2-divisible (and also $|X_{F_V}| = 1$).

Now assume that every real place of F has a 2-divisible value group. It suffices to prove $X_{F_1} = X_{F_1}$ in case F_1 is a finite algebraic extension of F . Let $v_1:\dot{F}_1 \to G_1$ be a real place of F_1 . Obviously v_1 extends some real place $v:\dot{F} \to G$ of F . Since $|G_1/G|$ is finite and G 2-divisible, G_1 is also 2-divisible. Thus F_1 satisfies (9.1)(c) which is euqivalent to $X_{F_1} = Y_{F_1}$.

q.e.d.

Algebraic Function Fields

The last corollary deals with fields whose real places have only 2-divisible value groups. Next let us look at fields admitting $\mathbb{Z}$ as value group of real places.

Let us take first $\mathbb{R}(x)$, the field of rational functions in x over $\mathbb{R}$. Let $v:\mathbb{R}(x) \to G$ be a real place of $\mathbb{R}(x)$. Then the restriction of v to $\mathbb{R}$ is trivial (i.e. $v(a) = 0$ for all $a \in \mathbb{R}$) or a real place of $\mathbb{R}$. But by (7.10), $\mathbb{R}$ cannot have a real place. Hence by (7.1), $G \simeq \mathbb{Z}$ and $(\mathbb{R}(x))_v \simeq \mathbb{R}$. Therefore the condition (c) of (9.1) is satisfied. Thus $\mathbb{R}(x)$ is one more example of an SAP-field.

How about $\mathbb{Q}(x)$ or $\mathbb{R}(y)(x)$? These fields are not SAP-fields. Since in case $\mathbb{Q}(x)$ the irreducible polynomial $x^2 - 2$ determines (by (7.1)) a real place with value group $\mathbb{Z}$ and residue class field $\mathbb{Q}(\sqrt{2})$, which admits two orderings. In case $\mathbb{R}(y)(x)$ the irreducible polynomial $x-1$ (over the coefficient field $\mathbb{R}(y)$) determines a real place with value group $\mathbb{Z}$ and residue class field $\mathbb{R}(y)$, which has in fact infinitely many orderings.

We are now going to work out the property of the field F_o of coefficients responsible for the field $F_o(x)$ to be SAP .

A field F is called <u>euclidean</u> if it is formally real and every element or its negative is a square. Hence F is euclidean iff F^2 is a positive cone of F . Every real closed field is euclidean.

F is called <u>hereditarily euclidean</u> if F and every formally real algebraic extension of F are euclidean. Every real closed field is hereditarily euclidean. These fields have been characterized by M. Ziegler and the author in [PZ].

(9.3) PROPOSITION F <u>is hereditarily euclidean iff</u> F <u>and every formally real algebraic extension of</u> F <u>admits exactly one ordering</u>.

<u>Proof</u>: To prove the non-trivial direction, suppose F_1 is algebraic over F and formally real, but not euclidean. Let P_1 be an ordering of F_1. There is $a \in \dot{P}_1$ which is not a square. Hence $F_1(\sqrt{a})$ admits at least two orderings.

q.e.d.

(9.4) THEOREM F_o <u>is hereditarily euclidean iff every formally real algebraic extension of</u> $F_o(x)$ <u>is an SAP-field</u>.

<u>Proof</u>: First let $F_o(x)$ be an SAP-field. Let $F_1 = F_o(\alpha)$ be a formally real finite algebraic extension of F_o and $f \in F_o[x]$ the minimal polynomial of α. Then by (7.1) f determines a real place of $F_o(x)$ with value group $\mathbb{Z}$ and residue class field F_1. Hence by (9.1)(c), F_1 has only one ordering. Thus F_o is hereditarily euclidean by (9.3).

Now assume F_o to be hereditarily euclidean. Let F be a finite algebraic formally real extension of $F_o(x)$. We prove condition (c) of (9.1) in order to show that F is an SAP-field. The case of an infinite algebraic formally real extension of $F_o(x)$ follows at once from the finite one (using e.g. condition (d) of (9.1)).

Let $v:\dot{F} \to G$ be a real place of F. The restriction v_o of v to F_o has 2-divisible value group G_o (compare (9.2)).

<u>Case 1</u>: G is contained in the divisible hull of G_o. Then G itself

is 2-divisible, and we are done.

Case 2: There is $f \in F_o(x)$ such that $v(f)$ is not an element of the divisible hull of G_o. Consider $F_o(f) \subset F_o(x)$. Let v^* be the restriction of v to $F_o(f)$. If $n \neq m$ and $a,b \in \dot{F}_o$, then obviously $v^*(af^n) \neq v^*(bf^m)$. Hence

$$v^*(\Sigma\, a_i\, f^i) = \min\{v_o(a_i) + iv(f)\} .$$

It is now easy to show that $v^*(F_o(f)) \simeq \mathbb{Z}\times G_o$ and $(F_o(f))_{v^*} \simeq F_{o\,v_o}$. Since $[F:F_o(f)]$ is finite, $|G/\mathbb{Z}\times G_o|$ and $[F_v:F_{o\,v_o}]$ are also finite. This implies first $|G/2G| = 2$, since G_o is 2-divisible. Next we will show that F_v has only one ordering. Once we have shown this, F will be an SAP-field by (9.1)(c). Let $F_v = F_{o\,v_o}(\alpha)$ and f a monic polynomial of $A_{v_o}[y]$ such that modulo M_{v_o} it constitutes the minimal polynomial of α over $F_{o\,v_o}$. Then f is irreducible over F_o as well. Let v_1 be an extension of v_o to $F_1 := F_o[y]/(f)$. Obviously $F_{1\,v_1} \simeq F_{o\,v_o}(\alpha) = F_v$. Since F_o is hereditarily euclidean and F_1 formally real (v_1 is a real place), F_1 admits exactly one ordering P_1. Hence $\{P_1\} = X_{F_1}^{v_1}$. Now by (7.9) there can only be one ordering on $F_{1\,v_1}$. Thus also $|X_{F_v}| = 1$.

q.e.d.

In [ELP] there is another proof of the fact that F_o hereditarily euclidean implies that every algebraic formally real extension K of $F_o(x)$ is SAP. Actually it is shown there that any quadratic form over K of dimension ≥ 3 which is indefinite with respect to all orderings is in fact isotropic over K. Obviously by (9.1)(e) this implies that K is SAP.

Looking at (9.4) the following question arises naturally. Is it possible that $F_o(x)$ is not SAP, but some finite algebraic formally real extension F_1 of $F_o(x)$ is SAP ? (For infinite algebraic extensions this is always possible. Simply take some real algebraic closure of $F_o(x)$. By (9.2) this is SAP.) More generally we may ask: if $F \subset F_1$ is a finite algebraic extension of formally real fields does SAP pass from one to the other field? One direction is not always valid. If F is SAP, F_1 need not be SAP. For example $F = \mathbb{Q}((\mathbb{Z}))$ is SAP, but $F_1 = F(\sqrt{2}) = \mathbb{Q}(\sqrt{2})((\mathbb{Z}))$ is not SAP. (Compare the examples to (8.8).)

About the other direction

$$F_1 \text{ is SAP} \Rightarrow F \text{ is SAP}$$

we don't know in general. Obviously the main problem is, what can we say about the extension of real places of F to real places of F_1. In case F (and hence also F_1) is an algebraic function field this problem can be answered satisfactorily. By an <u>algebraic function field</u> we mean a field F which is finitely generated but not algebraic over some subfield. We will prove the following theorem.

(9.5) THEOREM <u>Let</u> $F \subset F_1$ <u>be a finite algebraic field extension</u>. <u>Let</u> F_1 <u>be formally real and</u> F <u>be an algebraic function field. Then</u> F <u>is an SAP-field iff</u> F_1 <u>is an SAP-field</u>.

Before proving this theorem we have to look at the real places of algebraic function fields. First consider the field $F_o(x)$ of rational functions in x over F_o. By (7.1) the real places v of $F_o(x)$ <u>trivial on</u> F_o, i.e. $v(a) = 0$ for all $a \in F_o$, correspond to the irreducible polynomials $f \in F_o[x]$ such that $F_o[x]/(f)$ is formally real. Let P_o be some fixed ordering of F_o and R_o the real algebraic closure of $\langle F_o, P_o\rangle$. Then any $\alpha \in R_o$ determines a real place v_α of $F_o(x)$ trivial on F_o, simply by taking the minimal polynomial of α over F_o. The residue class field of v_α obviously is isomorphic to $F_o(\alpha)$. This isomorphism is given by $x + M_{v_\alpha} \to \alpha$.

(9.6) THEOREM <u>Let</u> $\langle F,P\rangle$ <u>be an ordered finite algebraic extension of</u> $F_o(x)$. <u>Let</u> R_o <u>be the real algebraic closure of</u> $\langle F_o, < \rangle$, <u>where</u> $<$ <u>is induced by</u> P . <u>Then there is</u> $\alpha \in R_o$ <u>such that</u> v_β <u>extends to a real place of</u> F <u>for all</u> $\beta \in R_o$ <u>in a sufficiently small neighbourhood of</u> α (i.e. $|\beta-\alpha| < \varepsilon$ <u>for some</u> $\varepsilon \in R_o$, $\varepsilon > 0$).

<u>Proof</u>: Let R_1 be the real algebraic closure of $\langle F,P\rangle$ and without loss of generality $R_o \subset R_1$. Let $f(X_1,X_2) \in R_o[X_1,X_2]$ be an irreducible polynomial such that $f(x,y) = 0$ for some generating element y of F over $F_o(x)$. Then obviously f satisfies the assumptions of Theorem (5.13). As a consequence of (5.13) we obtain the existence of $\alpha \in R_o$ and some neighborhood U of α in R_o such that $f(\beta,X_2)$ has a simple zero in R_o for all $\beta \in U$. Hence $f(\beta,X_2)$ changes sign in R_o (this follows directly from (3.4)). For every $\beta \in U$ there are $\gamma_1,\gamma_2 \in R_o$ such that

$$(*) \qquad f(\beta,\gamma_1) < 0 < f(\beta,\gamma_2).$$

Now the real place $v'_\beta : R_o(x) \to \mathbb{Z}$ determined by the polynomial $x - \beta$ obviously extends the real place v_β of $F_o(x)$. Choose some ordering P' of $R_o(x)$ such that v'_β is compatible with P'(use (7.10)). Let R' be the real algebraic closure of $\langle R_o(x),P'\rangle$. Then by (7.22), v'_β extends to R'. Its residue class field is R_o . Since the residue class of x is β, (*) implies

$$f(x,\gamma_1) < 0 < f(x,\gamma_2)$$

where $<$ now denotes the unique ordering of R'. But then by (3.4), $f(x,X_2)$ has a zero in R'. Thus F embeds into R'. This shows that v_β can be extended to a real place of F .

q.e.d.

(9.7) LEMMA <u>Let</u> $F_1 \subset F_2$ <u>be a finite algebraic extension of fields and</u> P_1 <u>an ordering of</u> F_1 <u>extending uniquely to</u> F_2. <u>Then every ordering of</u> F_1 <u>extends to</u> F_2 . <u>In particular, if</u> F_2 <u>bears a</u>

unique ordering, then F_1 also does.

Proof: By Corollary (3.12) the number of extensions of P_1 to F_2 equals the number of embeddings of F_2 into the real algebraic closure R of $<F_1,P_1>$ which is the identity on F_1. Let $F_2 = F_1(\alpha)$ and $f \in F_1[x]$ the minimal polynomial of α over F_2. Since P_1 extends uniquely, f has only one zero in R. But then by (3.4) the degree of f is odd. Now the conclusion follows from (1.26).

q.e.d.

The next theorem is a strengthening of (9.4) and will imply (9.5).

(9.8) THEOREM Let F be a finite algebraic formally real extension of $F_0(x)$. Then F is SAP iff F_0 is hereditarily euclidean.

Proof: If F_0 is hereditarily euclidean then by (9.4) F is SAP.

Now let F be SAP. Fix some ordering P of F. Let R_0 be the real algebraic closure of $<F_o, P \cap F_o>$. If F_0 is not hereditarily euclidean, then there is $\gamma \in R_0$ such that $F_0(\gamma)$ has more than one ordering (by (9.3)). The residue class field of v_γ is $F_0(\gamma)$. If we can extend v_γ to a real place of F, then (using Lemma (9.7)) the residue class field of this extension cannot have only one ordering. Since the value group of the extension is again isomorphic to $\mathbb{Z}$, (9.1)(c) implies a contradiction. Unfortunately v_γ itself may not extend. However, we will show that the set of $\gamma \in R_0$ such that $F_0(\gamma)$ has more than one ordering is dense in R_0. Now using Theorem (9.6) and the above argument we obtain a contradiction.

To show the above mentioned density let $\alpha \in R_0$. Consider a small neighbourhood determined by some $\varepsilon \in R_0$, $\varepsilon > 0$. Since F_0 is cofinal in R_0 (see (7.20)) we may even assume $\varepsilon \in F_0$. We will show that there is $\beta \in R_0$ such that $\alpha \le \beta < \alpha + \varepsilon$ and $F_0(\beta)$ has more than one ordering.

Let $F_o(\alpha,\gamma) = F_o(\delta)$ for some $\delta \in R_o$ (γ from above). By Lemma (9.7) $F_o(\delta)$ also has more than one ordering. Now consider

$$\beta \in \{\alpha + \frac{1}{n} \cdot \frac{\varepsilon}{1+\delta} \mid n \geq 1\} .$$

Obviously $F_o \subset F_o(\beta) \subset F_o(\delta)$. Since there are only finitely many fields in between, we get $F_o(\beta_1) = F_o(\beta_2)$ for some $\beta_i = \alpha + \frac{1}{n_i} \cdot \frac{\varepsilon}{1+\delta}$ such that $n_1 \neq n_2$. But then $F_o(\beta_1) = F_o(\delta)$. Hence $\alpha \leq \beta_1 < \alpha + \varepsilon$ and $F_o(\beta_1)$ has more than one ordering.

q.e.d.

Proof of (9.5): F is a finite algebraic extension of some rational function field $F_o(x)$. Now (9.5) follows from (9.8).

q.e.d.

From the extension Theorem (9.6) we now deduce an approximation theorem for the space X_F of orderings of a field F.

(9.9) THEOREM Let F be a finite algebraic formally real extension of $F_o(x)$ and $b_1,\ldots,b_m \in F$. Then the union of those X_F^v where v is a real place of F trivial on F_o and $v(b_1) = \ldots v(b_m) = 0$, is dense in X_F.

Proof: Let $P \in H_X(a_1) \cap \ldots \cap H_X(a_n)$ and $a_1,\ldots,a_n \in \dot{F}$. Then $a_1,\ldots,a_n \in P$. Hence by (1.26) P extends to $F_1 = F(\sqrt{a_1},\ldots,\sqrt{a_n})$. Let R_o be the real algebraic closure of $\langle F_o, P \cap F_o\rangle$. Then by (9.6) there is $\alpha \in R_o$ and a neighborhood U of α in R_o such that v_β extends to a real place of F_1 for all $\beta \in U$. It is well-known that there is only a finite number of $\gamma \in R_o$ such that $v(b_i) \neq 0$ for some $i \in \{1,\ldots,m\}$ and an extension v of v_γ to F. Choose $\beta \in U$ different from all these γ's. Let v' be an extension of v_β to a real place of F_1 and v its restriction to F. Then $v(b_1) = \ldots = v(b_m) = 0$. Next let $P' \in X_{F_1}^{v'}$. Since $a_1,\ldots,a_n$ are squares in F_1, $a_1,\ldots,a_n \in P' \cap F$. Hence $P' \cap F \in H_X(a_1) \cap \ldots \cap H_X(a_n)$ and $P' \cap F \in X_F^v$.

q.e.d.

§ 10. QUADRATIC FORMS OVER FORMALLY REAL FIELDS

(Part II)

In this paragraph we continue the study of quadratic forms over formally real fields from § 2. We will introduce the Witt ring $W(F)$ of a field F and study the connection between minimal prime ideals of $W(F)$ and orderings of F. Finally we present a characterization of the Witt rings corresponding to SAP-fields. Although there exists a large theory of Witt rings of fields (see [La]), we introduce only what we need to present the mentioned topics.

As in § 2 let F be a field of characteristic $\neq 2$. First we state without proof (for a proof see [La]) Witt's Cancellation Theorem

(10.1) THEOREM - Let ρ, ρ_1, ρ_2 be quadratic forms over F. Then $\rho \perp \rho_1 \simeq \rho \perp \rho_2$ implies $\rho_1 \simeq \rho_2$.

Recall that any quadratic form of dimension n is equivalent to some diagonalization $\langle a_1,\dots,a_n\rangle$. Regularity holds iff all a_i are $\neq 0$.

(10.2) PROPOSITION - A regular quadratic form ρ is isotropic iff $\rho \simeq \langle a,-a\rangle \perp \rho'$ for some ρ' and $a \in F$.

Proof: For the non-trivial direction let ρ be isotropic. Furthermore let $\rho \simeq \langle a_1,\dots,a_n\rangle$. Then $0 = \sum_{i=1}^{n} a_i v_i^2$ for some $v_i \in F$, and without loss of generality $v_1 \neq 0$. Hence $-a_1 = \sum_{i=2}^{n} a_i (\frac{v_i}{v_1})^2$. Therefore the $(n-1)$-dimensional form $\langle a_2,\dots,a_n\rangle$ represents $-a_1$. By Corollary (2.5) this implies $\langle a_2,\dots,a_n\rangle \simeq \langle -a_1,c_3,\dots,c_n\rangle$ for some $c_j \in F$. But then $\langle a_1,a_2,\dots,a_n\rangle \simeq \langle a_1,-a_1,c_3,\dots,c_n\rangle$.

q.e.d.

(10.3) LEMMA - Let ρ be a 2-dimensional regular quadratic form. Then (1) to (4) are equivalent:

(1) ρ <u>is isotropic</u>

(2) $\rho \simeq \langle a,-a\rangle$ <u>for some</u> $a \in \dot{F}$

(3) $\rho \simeq \langle 1,-1\rangle$

(4) $d(\rho) = -1\cdot\dot{F}^2$.

<u>Proof:</u> (1) $\Longleftrightarrow$ (2) follows from (10.2), and (3) $\Rightarrow$ (4) is trivial.
(4) $\Rightarrow$ (2): Let $\rho \simeq \langle c,d\rangle$. From (4) we get $d(\rho) = cd\cdot\dot{F}^2 = -1\cdot\dot{F}^2$.
Hence $d \equiv -c \bmod \dot{F}^2$. Thus $\rho \simeq \langle c,-c\rangle$.
(2) $\Rightarrow$ (3): Since $1 = a(\frac{1+a}{2a})^2 - a(\frac{1-a}{2a})^2$, by Corollary (2.5),
$\rho \simeq \langle 1,c\rangle$. From $d(\rho) = -a^2\cdot\dot{F}^2 = c\cdot\dot{F}^2$ we obtain $c \equiv -1 \bmod \dot{F}^2$.
Hence $\rho \simeq \langle 1,-1\rangle$.

q.e.d.

The 2-dimensional form $\langle 1,-1\rangle$ is called the <u>hyperbolic form</u> or <u>plane</u>. A 2n-dimensional form ρ is called <u>hyperbolic</u> if $\rho \simeq n\langle 1,-1\rangle$.

(10.4) THEOREM (Witt [W]) - <u>Every quadratic form</u> ρ <u>over</u> F <u>admits a decomposition</u>

$$\rho \simeq r\langle 0\rangle \perp s\langle 1,-1\rangle \perp \rho_a ,$$

<u>where</u> ρ_a <u>is anisotropic. The natural numbers</u> r <u>and</u> s <u>and the equivalence class of</u> ρ_a <u>are uniquely determined by</u> ρ .

<u>Proof:</u> For the existence let $\rho \simeq r\langle 0\rangle \perp \rho_1$ such that ρ_1 is regular. Now apply Proposition (10.2) till the desired form is reached. To prove the uniqueness use Witt's Cancellation Theorem (10.1).

q.e.d.

ρ_a will be called the <u>anisotropic part</u> of ρ .

We now introduce a new equivalence relation on the set of <u>regular</u> quadratic forms (of arbitrary dimension) over F . Let ρ_1 and ρ_2 be regular. Then ρ_1 is called <u>similar</u> to ρ_2 (we write $\rho_1 \sim \rho_2$) if

$$\rho_1 \perp n\langle 1,-1\rangle \simeq \rho_2 \perp m\langle 1,-1\rangle$$

for some $n,m \in \mathbb{N}$, i.e. up to some hyperbolic planes they are equivalent. From (10.1) trivially follows:

$$(10.5) \qquad \rho_1 \simeq \rho_2 \quad \text{iff} \begin{cases} \rho_1 \sim \rho_2 & \text{and} \\ \dim \rho_1 = \dim \rho_2 & . \end{cases}$$

Obviously $\perp$ and $\otimes$ induce corresponding operations on the similarity classes of regular quadratic forms.

(10.6) LEMMA <u>The set of similarity classes of regular quadratic forms together with the operations induced by $\perp$ and $\otimes$ form a commutative ring.</u>

<u>Proof</u>: The Lemma follows from the remark on p. 19 and

$$\begin{aligned} \langle a_1,\dots,a_n\rangle \perp \langle -a_1,\dots,-a_n\rangle &\simeq \langle a_1,-a_1\rangle \perp\dots\perp \langle a_n,a_n\rangle \\ &\simeq \langle 1,-1\rangle \perp\dots\perp\langle 1,-1\rangle \\ &\sim 0 \end{aligned}$$

q.e.d.

This ring is called the <u>Witt ring</u> $W(F)$ of the field F. Note that $W(F)$ is additively generated by the similarity classes of $\langle a\rangle$ for $a \in \dot{F}$.

(10.7) PROPOSITION <u>Every similarity class contains exactly one equivalence class of anisotropic quadratic forms.</u>

<u>Proof</u>: From $\rho' \sim \rho''$ we deduce $\rho'_a \sim \rho''_a$. Hence $\rho'_a \perp n\langle 1,-1\rangle \simeq$ $\simeq \rho''_a \perp m\langle 1,-1\rangle$ for some $n,m \in \mathbb{N}$. But necessarily $n = m$, and hence $\rho'_a \simeq \rho''_a$.

q.e.d.

(10.8) EXAMPLES

(1) Let $F = \mathbb{C}$. Then every element of F is a square and hence $\langle a_1,\dots,a_n\rangle \simeq \langle 1,\dots,1\rangle$ (if all $a_i \neq 0$). In particular $\langle 1,-1\rangle \simeq \langle 1,1\rangle$. Hence

$$\rho \sim \begin{cases} 0 & \text{if } \dim\rho \text{ is even} \\ \langle 1\rangle & \text{if } \dim\rho \text{ is odd.} \end{cases}$$

Therefore $W(\mathbb{C}) \simeq \mathbb{Z}/2\mathbb{Z}$.

(2) Let $F = \mathbb{R}$. Then $\dot{F} = \dot{F}^2 \cup (-1)\dot{F}^2$ and hence $<a_1,\ldots,a_n> \simeq <1,\ldots,1,-1,\ldots,-1>$ (if all $a_i \neq 0$).
Let sgn denote the signature of quadratic forms with respect to the unique ordering of $\mathbb{R}$. Since $\mathrm{sgn} <1,-1> = 0$, obviously sgn induces a homomorphism $\mathrm{sgn}: W(F) \to \mathbb{Z}$. If $\mathrm{sgn}\, \rho = 0$, then $\rho \simeq m<1> \perp m<-1> \simeq$ $\simeq m<1,-1> \sim 0$. Hence sgn is an isomorphism between $W(F)$ and $\mathbb{Z}$.

Note that over $\mathbb{R}$ a quadratic form ρ is isotropic iff $|\mathrm{sgn}\, \rho| < \dim \rho$, and ρ is hyperbolic iff $\mathrm{sgn}\, \rho = 0$. Note also all of Example (2) still holds in F is euclidean.

Next we will consider the prime ideals of $W(F)$. By I we denote the ideal consisting of similarity classes belonging to even-dimensional forms. This ideal is also called the <u>fundamental ideal</u> of $W(F)$.

(10.9) PROPOSITION <u>Let</u> $\mathfrak{q}$ <u>be a prime ideal of</u> $W(F)$. <u>Then</u> $W(F)/\mathfrak{q} \simeq \mathbb{Z}$ <u>or</u> $\mathbb{Z}/p\mathbb{Z}$ <u>for some prime number</u> $p \in \mathbb{N}$. I <u>is a maximal ideal in</u> $W(F)$, <u>and</u> $\mathfrak{q} = I$ iff $<1,1> \in \mathfrak{q}$.

<u>Proof</u>: Let $\mathfrak{q}$ be a prime ideal of $W(F)$. Then
$0 \sim (<a^2>\perp<-1>) \sim (<a>\perp<1>) \otimes (<a>\perp<-1>) \in \mathfrak{q}$ implies $<a> \equiv <\pm 1> \bmod \mathfrak{q}$. Hence the map $n \to n<1>$ induces a homomorphism from $\mathbb{Z}$ onto $W(F)/\mathfrak{q}$. Thus $W(F)/\mathfrak{q} \simeq \mathbb{Z}$ or $\mathbb{Z}/p\mathbb{Z}$ for some prime number $p \in \mathbb{N}$. The maximality of I follows from $W(F)/I \simeq \mathbb{Z}/2\mathbb{Z}$. Since $<1,1> \in \mathfrak{q}$ implies $<1> \equiv <-1> \bmod \mathfrak{q}$, we obtain $<a> \equiv <1> \bmod \mathfrak{q}$ for all $a \in \dot{F}$. Hence $<1,1> \in \mathfrak{q}$ also implies $I \subset \mathfrak{q}$, thus $I = \mathfrak{q}$.

q.e.d.

Note that in commutative rings over every ideal there is a maximal one, which then also is a prime ideal. Moreover, below every prime ideal there is a minimal prime ideal.

(10.10) THEOREM If the field F is not formally real, then I is the only prime ideal of $W(F)$.

Proof: Let $\mathfrak{q}$ be a prime ideal of $W(F)$ such that $\langle 1,1\rangle \notin \mathfrak{q}$. Consider

$$P := \{a \in \dot{F} \mid \langle a\rangle \equiv \langle 1\rangle \mod \mathfrak{q}\}.$$

Obviously $P\cdot P \subset P$ and $P \cup -P = \dot{F}$. Moreover $P+P \subset P$. To prove this let $\langle a\rangle \equiv \langle 1\rangle$ and $\langle b\rangle \equiv \langle 1\rangle \mod \mathfrak{q}$. Obviously $a+b \neq 0$, since $\langle 1\rangle \not\equiv \langle -1\rangle \mod \mathfrak{q}$. Therefore by Corollary (2.5), $\langle a,b\rangle \simeq \langle a+b,x\rangle$ for some $x \in \dot{F}$. Computing determinants we obtain $ab\cdot\dot{F}^2=(a+b)x\cdot\dot{F}^2$, implying $x \equiv (a+b)ab \mod \dot{F}^2$. Hence $\langle a,b\rangle \simeq \langle a+b, (a+b)ab\rangle$. Now assume $\langle a+b\rangle \equiv \langle -1\rangle \mod \mathfrak{q}$. Then $2\langle 1\rangle \equiv \langle a\rangle \perp \langle b\rangle \equiv \langle a+b\rangle \perp \langle (a+b)ab\rangle \equiv$ $\equiv \langle -1\rangle \perp \langle -1\rangle \mod \mathfrak{q}$ implies $4\langle 1\rangle \equiv 0 \mod \mathfrak{q}$. But $4\langle 1\rangle \simeq 2\langle 1\rangle \otimes 2\langle 1\rangle$. Thus we would get the contradiction $2\langle 1\rangle \in \mathfrak{q}$. This proves $P+P \subset P$.

But then P is a positive cone of F (see page 62). Since F is not formally real, by this argument there can be no prime ideal $\mathfrak{q}$ of F such that $\langle 1,1\rangle \notin \mathfrak{q}$.

q.e.d.

Recall, if $P \in X_F$ is an ordering of the field F, then $\mathrm{sgn}_P\,\rho$ denotes the signature of the quadratic form ρ with respect to P. The signature is invariant under equivalence of quadratic forms (Theorem (2.6)) and obviously $\mathrm{sgn}_P\langle 1,-1\rangle = 0$. Hence the signature induces a homomorphism

$$\mathrm{sgn}_P\colon W(F) \to \mathbb{Z}.$$

(10.11) THEOREM If F is formally real, the minimal prime ideals of $W(F)$ correspond uniquely to the orderings of F. The correspondence is given by

$$P \to \mathrm{kernel}(\mathrm{sgn}_P).$$

Proof: Obviously for every ordering P of F, $\mathrm{kernel}(\mathrm{sgn}_P)$ is a prime ideal $\mathfrak{q}$ such that $\langle 1,1\rangle \notin \mathfrak{q} \subset I$. Hence I is not a minimal prime ideal.

Now let $\mathfrak{q}$ be a prime ideal such that $<1,1> \notin \mathfrak{q}$. Then the proof of (10.10) shows that $P = \{a \in \dot{F} \mid <a> \equiv <1> \bmod \mathfrak{q}\}$ is a positive cone of F. Clearly $\text{kernel}(\text{sgn}_P) \subset \mathfrak{q}$. Hence any minimal prime ideal of $W(F)$ is the kernel of some signature.

Note that $\text{kernel}(\text{sgn}_{P_1}) \subset \text{kernel}(\text{sgn}_{P_2})$ implies $P_1 \subset P_2$ and hence $P_1 = P_2$, but then also $\text{kernel}(\text{sgn}_{P_1}) = \text{kernel}(\text{sgn}_{P_2})$. To show this, let $a \in P_1$. Then $\text{sgn}_{P_1} <a,-1> = 0$. Hence $\text{sgn}_{P_2} <a,-1> = 0$, proving $a \in P_2$. This remark proves that $\text{kernel}(\text{sgn}_P)$ is minimal and that the correspondence in the theorem is one-to-one.

q.e.d.

Let us now introduce the <u>total signature</u> $\text{sgn}\,\rho$ of a quadratic form ρ by

$$\begin{aligned} \text{sgn}\,\rho : X_F &\to \mathbb{Z} \\ P &\to \text{sgn}_P\,\rho . \end{aligned}$$

In §6 we introduced a topology on X_F. Using this topology we get:

(10.12) PROPOSITION <u>The total signature</u> $\text{sgn}\,\rho$ <u>is a continuous map, if</u> $\mathbb{Z}$ <u>is endowed with the discrete topology</u>.

<u>Proof</u>: This follows trivially from $\text{sgn}<a>^{-1}(\pm 1) = H(\pm a)$ and $\text{sgn} < a_1,\dots,a_n> = \text{sgn} < a_1> + \dots + \text{sgn} < a_n>$.

q.e.d.

Now let $C(X_F,\mathbb{Z})$ denote the ring of continuous maps from X_F to $\mathbb{Z}$. Then the total signature obviously induces a homomorphism

$$\text{sgn}: W(F) \to C(X_F,\mathbb{Z})$$

by $\rho \mapsto \text{sgn}\,\rho$. For the kernel of this homomorphism we obtain:

$$\text{kernel(sgn)} = \bigcap_{P \in X_F} \text{kernel}(\text{sgn}_P) = \bigcap_{\mathfrak{q} \text{ min.prime}} \mathfrak{q} = \bigcap_{\mathfrak{q} \text{ prime}} \mathfrak{q}$$

$$= \text{Nil } W(F) \text{ (= nil radical of } W(F)).$$

It will turn out that Nil W(F) consists exactly of the torsion elements (of the additive group) of W(F). Let us first consider the non-real case.

(10.13) LEMMA <u>If</u> F <u>is not formally real, then</u> Nil W(F) = I <u>and</u> W(F) <u>has a</u> 2^n-<u>torsion for some</u> $n \in \mathbb{N}$.

<u>Proof</u>: Nil W(F) = I is implied by (10.10). Now $<1,1> \in I$ implies $<1,1> \otimes \ldots \otimes <1,1> \sim 0$ for some product. Hence $2^n<1> \sim 0$ for some $n \in \mathbb{N}$.

q.e.d.

To treat the real case we need a lemma.

(10.14) LEMMA <u>Let</u> ρ <u>be anisotropic in</u> F <u>and hyperbolic in</u> $F(\sqrt{d})$. <u>Then</u> $\rho \simeq <1,-d> \otimes \rho'$ <u>for some</u> ρ'. <u>In particular</u> $\rho \simeq <-d> \otimes \rho$.

<u>Proof</u>: Since ρ is hyperbolic in $F(\sqrt{d})$ it is also isotropic. Let $\rho \simeq <a_1,\ldots,a_n>$. Then $0 = \sum_{i=1}^{n} a_i(v_i + w_i\sqrt{d})^2$ for some $v_i, w_i \in F$, not all equal to 0 . Since ρ is not isotropic in F , this implies $0 \neq \Sigma a_i v_i^2 = -d \Sigma a_i w_i^2$ and $\Sigma a_i v_i w_i = 0$. Now consider the inner product space (F^n, b_ρ) (where b_ρ is defined using the canonical base of F^n). Then we get $0 \neq b_\rho(\bar{v},\bar{v}) = -d\, b_\rho(\bar{w},\bar{w})$ and $b_\rho(\bar{v},\bar{w}) = 0$. Now from the proof of the Theorem (2.3) it follows that $\{\bar{w},\bar{v}\}$ can be extended to an orthogonal base of (F^n, b_ρ). Hence, letting

$c_1 := b_\rho(\bar{w},\bar{w})$, we obtain $\rho \simeq \langle c_1, -d\, c_1, \ldots\rangle \simeq \langle 1,-d\rangle \otimes \langle c_1\rangle \perp \rho_1$ for some ρ_1. Now ρ_1 is still anisotropic in F and hyperbolic in $F(\sqrt{d})$, since ρ and $\langle 1,-d\rangle$ are hyperbolic in $F(\sqrt{d})$. Hence by induction we get $\rho \simeq \langle 1,-d\rangle \otimes \langle c_1,\ldots,c_m\rangle$ for some $c_1,\ldots,c_m \in \dot{F}$. This proves the first assertion.

From $\rho \simeq \langle 1,-d\rangle \otimes \rho'$ we conclude $(\langle 1\rangle \perp \langle d\rangle) \otimes \rho \simeq (\langle 1\rangle \perp \langle d\rangle) \otimes$ $\otimes\ (\langle 1\rangle \perp \langle -d\rangle) \otimes \rho' \sim 0$. Hence $\langle 1\rangle\rho \simeq \langle -d\rangle\, \rho$.

q.e.d.

(10.15) THEOREM (Pfister [Pf]) <u>Let</u> F <u>be formally real and</u> ρ <u>be regular. Then</u> (1) <u>to</u> (5) <u>are euqivalent</u>:

(1) ρ <u>is nilpotent in</u> $W(F)$

(2) $\mathrm{sgn}_P\, \rho = 0$ <u>for all</u> $P \in X_F$

(3) ρ <u>is hyperbolic in all real closures of</u> F

(4) $2^n\, \rho \sim 0$ <u>for some</u> $n \in \mathbb{N}$

(5) $m\rho \sim 0$ <u>for some</u> $m \in \mathbb{N}$.

<u>Proof</u>: (1) $\Leftrightarrow$ (2) $\Leftrightarrow$ (3) is obvious, since $\mathrm{Nil}\, W(F) = \mathrm{kernel}(\mathrm{sgn})$. Also (4) $\Rightarrow$ (5) $\Rightarrow$ (2) is trivial.

(1) $\Rightarrow$ (4): Let ρ be nilpotent in $W(F)$. Assume $2^n\, \rho \not\sim 0$ for all $n \in \mathbb{N}$. Now let F_1 be a maximal extension of F (in a fixed algebraic closure) such that $2^n\, \rho \not\sim 0$ in $W(F_1)$ for all $n \in \mathbb{N}$. By (10.13), F_1 then is formally real. By Example (10.8)(2), F_1 cannot be euclidean, since ρ nilpotent implies $\mathrm{sgn}_P\, \rho = 0$. But this is equivalent to $\rho \sim 0$ for a euclidean field. Hence there are $a,b \in F_1$ such that 1, a, b and ab are different mod $\dot{F}_1^2$. By the maximality of F_1 it follows that there is some $m \in \mathbb{N}$ such that $2^m\, \rho \sim 0$ in $W(F_1(\sqrt{d}))$ for $d = a$, b and ab.

Let $\rho' := 2^m\, \rho$ and ρ'_a be its anisotropic part. By Lemma (10.14) we obtain $\rho'_a \simeq \langle -d\rangle \otimes \rho'_a$ for $d = a$, b and ab. This obviously also holds for the hyperbolic part (see (10.3)). Hence

$$\begin{aligned}\rho' &\simeq \langle -a\rangle \otimes \rho' \\ &\simeq \langle -b\rangle \otimes \rho' \\ &\simeq \langle -ab\rangle \otimes \rho'.\end{aligned}$$

This implies $\rho' \simeq \langle -ab\rangle \otimes \rho' \simeq \langle a\rangle \otimes \rho' \simeq \langle -1\rangle \otimes \rho'$. Hence $(\langle 1\rangle \perp \langle 1\rangle) \otimes \rho' \sim 0$. But then $2^{m+1}\ \rho \sim 0$, which is a contradiction to the choice of F_1 .

q.e.d.

By Theorem (10.15) the kernel of the total signature is known very well. Now, what can we say about the image? Obviously we have an embedding of the <u>reduced</u> Witt ring into $C(X_F,\mathbb{Z})$:

$$W(F)_{red} = W(F)/\mathrm{Nil}\ W(F) \to C(X_F,\mathbb{Z}).$$

Since $\mathrm{sgn}_P\ \rho \equiv \dim \rho \bmod (2)$ for all orderings, $W(F)_{red}$ embeds into the subring $\mathbb{Z}\cdot 1 + C(X_F,2\mathbb{Z})$ of $C(X_F,\mathbb{Z})$, where 1 denotes the function with constant value 1 .

The last theorem of this paragraph is due to Knebusch, Rosenberg and Ware (see [KRW]). It gives a characterization of Witt rings of SAP-fields.

(10.16) THEOREM F <u>is an</u> SAP-<u>field iff the image of the total signature coincides with the subring</u> $\mathbb{Z}\cdot 1 + C(X_F,2\mathbb{Z})$ <u>of</u> $C(X_F,\mathbb{Z})$.

<u>Proof</u>: First let F be an SAP-field. Since 1 is the image of $\langle 1\rangle$, it suffices to consider $f \in C(X_F,2\mathbb{Z})$. By the compactness of X_F , the inverse images $f^{-1}(2n)$ give a finite partition of X_F into clopen (closed and open) subsets. By Theorem (9.1) for every $f^{-1}(2n) \neq \emptyset$ there is $a_n \in \dot{F}$ such that $f^{-1}(2n) = H(a_n)$. Let $f_n := \mathrm{sgn}\ n\langle 1,a_n\rangle$. Obviously $f_n^{-1}(2n) = H(a_n)$ and f_n is zero outside of $H(a_n)$. Hence $f = \sum_n f_n = \mathrm{sgn}\ \sum_n n\langle 1,a_n\rangle$.

Now let $\text{sgn}(W(F)) = \mathbb{Z}\cdot 1 + C(X_F, 2\mathbb{Z})$. If $A \subset X_F$ is clopen, then the characteristic function χ_A of A, defined by $\chi_A(P) = 2$ if $P \in A$ and 0 if $P \notin A$, is an element of $C(X_F, 2\mathbb{Z})$. Hence there is a quadratic form ρ such that $\text{sgn}\,\rho = \chi_A$. Let $\rho = \langle a_1, \ldots, a_{2n}\rangle$.

<u>Case 1</u>: n even. Then $\det\rho \in P$ iff $\text{sgn}_P\,\rho = 0$ iff $P \notin A$. Hence $A = H(-\det\rho)$.

<u>Case 2</u>: n odd. Then $\det\rho \in P$ iff $\text{sgn}_P\,\rho = 2$ iff $P \in A$. Hence $A = H(\det\rho)$.

By Theorem (9.1) this proves F to be an SAP-field.

q.e.d.

More about the image of sgn in general can be found in L. Bröcker's paper [Br].

REFERENCES

[A] E. Artin: Über die Zerlegung definiter Funktionen in Quadrate. Abh.Math.Sem.Univ.Hamburg 5 (1927), 100-115.

[AS] E. Artin and O.Schreier: Algebraische Konstruktion reeller Körper. Abh.Math.Sem.Univ.Hamburg 5 (1927), 85-99.

[Ax] J. Ax: A Metamathematical Approach to Some Problems in Number Theory. AMS, SYMP.Vol.XX, 161-190.

[BS] E. Becker and K.-J.Spitzlay: Zum Satz von Artin-Schreier über die Eindeutigkeit des reellen Abschlusses eines angeordneten Körpers. Comment.Math.Helvetici 50 (1975), 81-87.

[B Sl] I.L. Bell and A.B.Slomson: Models and Ultraproducts. North Holland, 1971.

[B] W. Berger: Zur Theorie der Anordnungen eines Körpers. Diplomarbeit, Bonn 1975.

[Br] L. Bröcker: Zur Theorie der quadratischen Formen über formal reellen Körpern. Math.Ann. 210 (1974), 233-256.

[Ch K] C. Chang and J.Keisler: Model Theory. North Holland, 1973.

[Cr] T.C. Craven: The Boolean Space of Orderings of a Field. Trans.Amer.Math.Soc. 209 (1975), 225-235.

[Du] D.W. Dubois: A Nullstellensatz for Ordered Fields. Ark. Mat. 8 (1969), 111-114.

[Du E] D.W. Dubois and G.A.Efroymson: Algebraic Theory of Real Varieties. I. Studies and Essays Presented to Yu-Why-Chen on his Sixtieth Birthday, October 1970.

[El] R. Elman and T.Y.Lam: Quadratic forms over formally real fields and pythagorean fields. Amer.J.Math.94(1972),1155-1194.

[ELP] R.Elman,T.Y.Lam and A.Prestel: On Some Hasse Principles over Formally Real Fields. Math.Z. 134 (1973), 291-301.

[E] O. Endler: Valuation Theory. Springer, Berlin-Heidelberg-New York, 1972.

[Fu] L. Fuchs: Teilweise geordnete algebraische Strukturen. Vandenhoeck & Ruprecht, 1966.

[H] D.K. Harrison: Witt Rings. Lecture Notes. Department of Mathematics, University of Kentucky, Lexington, Kentucky, 1970.

[Kn] M. Knebusch: On the Uniqueness of Real Closures and the Existence of Real Places. Comment.Math.Helvetici 47 (1972), 260-269.

[KRW] M. Knebusch,A.Rosenberg and R.Ware: Structure of Witt Rings, Quotients of Abelian Group Rings, and Orderings of Fields. Bull.Amer.Math.Soc. 77 (1970), 205-209.

[Kr] W. Krull: Allgemeine Bewertungstheorie. J. reine angew.Math. 167 (1932), 160-196.

[La] T.Y. Lam: The Algebraic Theory of Quadratic Forms. Benjamin 1973.

[L] S. Lang: The Theory of Real Places. Annals of Math. 57 (1953), 378-391.

[LL] J. Leicht and F.Lorenz: Die Primideale des Wittschen Ringes. Invent. Math. 10 (1970), 378-391.

[MH] J. Milnor and D.Husemoller: Symmetric Bilinear Forms. Springer 1973.

[Pf] A. Pfister: Quadratische Formen in beliebigen Körpern. Invent.Math. 1 (1966), 116-132.

[P] A. Prestel: Quadratische Semi-Ordnungen und quadratische Formen. Math.Z. 133 (1973), 319-342.

[PZ] A. Prestel and M.Ziegler: Erblich euklidische Körper. J. reine angew. Math. 274/275 (1975), 196-205.

[Ri] P. Ribenboim: Théorie des Valuations. 2e edition. Les Presses de l'Université de Montréal, 1968.

[Sp] T.A. Springer: Sur les formes quadratiques d'indice zero. C.R. Acad.Sci. 234 (1952), 1517-1519.

[T] A. Tarski: The Completeness of Elementary Algebra and Geometry. Herman Cie, Paris, 1940.

[W] E. Witt: Theorie der quadratischen Formen in beliebigen Körpern. J. reine angew. Math. 176 (1937), 31-44.

Notations and Conventions

Unless stated differently, all fields are assumed to have characteristic $\neq 2$.

$A \setminus B \;=\; \{x \mid x \in A,\ x \notin B\}$

$\dot{A} \;=\; A \setminus \{0\}$

$A^2 \;=\; \{a^2 \mid a \in A\}$

$-A \;=\; \{-a \mid a \in A\}$

$A + B \;=\; \{a + b \mid a \in A,\ b \in B\}$

$A \cdot B \;=\; \{ab \mid a \in A,\ b \in B\}$

$|A| \;=$ number of elements of A

$S_A \;=\; \{\sum_{i=1}^{n} a_i^2 \mid n \in \mathbb{N},\ a_i \in A\}$

$\bar{g} \;=\; g + 2G$

INDEX

Vol. 952: Combinatorial Mathematics IX. Proceedings, 1981. Edited by E. Billington, S. Oates-Williams, and A.P. Street. XI, 443 pages. 1982.

Vol. 953: Iterative Solution of Nonlinear Systems of Equations. Proceedings, 1982. Edited by R. Ansorge, Th. Meis, and W. Törnig. VII, 202 pages. 1982.

Vol. 954: S.G. Pandit, S.G. Deo, Differential Systems Involving Impulses. VII, 102 pages. 1982.

Vol. 955: G. Gierz, Bundles of Topological Vector Spaces and Their Duality. IV, 296 pages. 1982.

Vol. 956: Group Actions and Vector Fields. Proceedings, 1981. Edited by J.B. Carrell. V, 144 pages. 1982.

Vol. 957: Differential Equations. Proceedings, 1981. Edited by D.G. de Figueiredo. VIII, 301 pages. 1982.

Vol. 958: F.R. Beyl, J. Tappe, Group Extensions, Representations, and the Schur Multiplicator. IV, 278 pages. 1982.

Vol. 959: Géométrie Algébrique Réelle et Formes Quadratiques, Proceedings, 1981. Edité par J.-L. Colliot-Thélène, M. Coste, L. Mahé, et M.-F. Roy. X, 458 pages. 1982.

Vol. 960: Multigrid Methods. Proceedings, 1981. Edited by W. Hackbusch and U. Trottenberg. VII, 652 pages. 1982.

Vol. 961: Algebraic Geometry. Proceedings, 1981. Edited by J.M. Aroca, R. Buchweitz, M. Giusti, and M. Merle. X, 500 pages. 1982.

Vol. 962: Category Theory. Proceedings, 1981. Edited by K.H. Kamps D. Pumplün, and W. Tholen, XV, 322 pages. 1982.

Vol. 963: R. Nottrot, Optimal Processes on Manifolds. VI, 124 pages. 1982.

Vol. 964: Ordinary and Partial Differential Equations. Proceedings, 1982. Edited by W.N. Everitt and B.D. Sleeman. XVIII, 726 pages. 1982.

Vol. 965: Topics in Numerical Analysis. Proceedings, 1981. Edited by P.R. Turner. IX, 202 pages. 1982.

Vol. 966: Algebraic K-Theory. Proceedings, 1980, Part I. Edited by R.K. Dennis. VIII, 407 pages. 1982.

Vol. 967: Algebraic K-Theory. Proceedings, 1980. Part II. VIII, 409 pages. 1982.

Vol. 968: Numerical Integration of Differential Equations and Large Linear Systems. Proceedings, 1980. Edited by J. Hinze. VI, 412 pages. 1982.

Vol. 969: Combinatorial Theory. Proceedings, 1982. Edited by D. Jungnickel and K. Vedder. V, 326 pages. 1982.

Vol. 970: Twistor Geometry and Non-Linear Systems. Proceedings, 1980. Edited by H.-D. Doebner and T.D. Palev. V, 216 pages. 1982.

Vol. 971: Kleinian Groups and Related Topics. Proceedings, 1981. Edited by D.M. Gallo and R.M. Porter. V, 117 pages. 1983.

Vol. 972: Nonlinear Filtering and Stochastic Control. Proceedings, 1981. Edited by S.K. Mitter and A. Moro. VIII, 297 pages. 1983.

Vol. 973: Matrix Pencils. Proceedings, 1982. Edited by B. Kågström and A. Ruhe. XI, 293 pages. 1983.

Vol. 974: A. Draux, Polynômes Orthogonaux Formels – Applications. VI, 625 pages. 1983.

Vol. 975: Radical Banach Algebras and Automatic Continuity. Proceedings, 1981. Edited by J.M. Bachar, W.G. Bade, P.C. Curtis Jr., H.G. Dales and M.P. Thomas. VIII, 470 pages. 1983.

Vol. 976: X. Fernique, P.W. Millar, D.W. Stroock, M. Weber, Ecole d'Eté de Probabilités de Saint-Flour XI – 1981. Edited by P.L. Hennequin. XI, 465 pages. 1983.

Vol. 977: T. Parthasarathy, On Global Univalence Theorems. VIII, 106 pages. 1983.

Vol. 978: J. Ławrynowicz, J. Krzyż, Quasiconformal Mappings in the Plane. VI, 177 pages. 1983.

Vol. 979: Mathematical Theories of Optimization. Proceedings, 1981. Edited by J.P. Cecconi and T. Zolezzi. V, 268 pages. 1983.

Vol. 980: L. Breen. Fonctions thêta et théorème du cube. XIII, 115 pages. 1983.

Vol. 981: Value Distribution Theory. Proceedings, 1981. Edited by I. Laine and S. Rickman. VIII, 245 pages. 1983.

Vol. 982: Stability Problems for Stochastic Models. Proceedings, 1982. Edited by V. V. Kalashnikov and V. M. Zolotarev. XVII, 295 pages. 1983.

Vol. 983: Nonstandard Analysis-Recent Developments. Edited by A.E. Hurd. V, 213 pages. 1983.

Vol. 984: A. Bove, J. E. Lewis, C. Parenti, Propagation of Singularities for Fuchsian Operators. IV, 161 pages. 1983.

Vol. 985: Asymptotic Analysis II. Edited by F. Verhulst. VI, 497 pages. 1983.

Vol. 986: Séminaire de Probabilités XVII 1981/82. Proceedings. Edited by J. Azéma and M. Yor. V, 512 pages. 1983.

Vol. 987: C. J. Bushnell, A. Fröhlich, Gauss Sums and p-adic Division Algebras. XI, 187 pages. 1983.

Vol. 988: J. Schwermer, Kohomologie arithmetisch definierter Gruppen und Eisensteinreihen. III, 170 pages. 1983.

Vol. 989: A.B. Mingarelli, Volterra-Stieltjes Integral Equations and Generalized Ordinary Differential Expressions. XIV, 318 pages. 1983.

Vol. 990: Probability in Banach Spaces IV. Proceedings, 1982. Edited by A. Beck and K. Jacobs. V, 234 pages. 1983.

Vol. 991: Banach Space Theory and its Applications. Proceedings, 1981. Edited by A. Pietsch, N. Popa and I. Singer. X, 302 pages. 1983.

Vol. 992: Harmonic Analysis, Proceedings, 1982. Edited by G. Mauceri, F. Ricci and G. Weiss. X, 449 pages. 1983.

Vol. 993: R.D. Bourgin, Geometric Aspects of Convex Sets with the Radon-Nikodým Property. XII, 474 pages. 1983.

Vol. 994: J.-L. Journé, Calderón-Zygmund Operators, Pseudo-Differential Operators and the Cauchy Integral of Calderón. VI, 129 pages. 1983.

Vol. 995: Banach Spaces, Harmonic Analysis, and Probability Theory. Proceedings, 1980–1981. Edited by R.C. Blei and S.J. Sidney. V, 173 pages. 1983.

Vol. 996: Invariant Theory. Proceedings, 1982. Edited by F. Gherardelli. V, 159 pages. 1983.

Vol. 997: Algebraic Geometry – Open Problems. Edited by C. Ciliberto, F. Ghione and F. Orecchia. VIII, 411 pages. 1983.

Vol. 998: Recent Developments in the Algebraic, Analytical, and Topological Theory of Semigroups. Proceedings, 1981. Edited by K.H. Hofmann, H. Jürgensen and H. J. Weinert. VI, 486 pages. 1983.

Vol. 999: C. Preston, Iterates of Maps on an Interval. VII, 205 pages. 1983.

Vol. 1000: H. Hopf, Differential Geometry in the Large, VII, 184 pages. 1983.

Vol. 1001: D.A. Hejhal, The Selberg Trace Formula for PSL(2, ℝ). Volume 2. VIII, 806 pages. 1983.

Vol. 1002: A. Edrei, E.B. Saff, R.S. Varga, Zeros of Sections of Power Series. VIII, 115 pages. 1983.

Vol. 1003: J. Schmets, Spaces of Vector-Valued Continuous Functions. VI, 117 pages. 1983.

Vol. 1004: Universal Algebra and Lattice Theory. Proceedings, 1982. Edited by R.S. Freese and O.C. Garcia. VI, 308 pages. 1983.

Vol. 1005: Numerical Methods. Proceedings, 1982. Edited by V. Pereyra and A. Reinoza. V, 296 pages. 1983.

Vol. 1006: Abelian Group Theory. Proceedings, 1982/83. Edited by R. Göbel, L. Lady and A. Mader. XVI, 771 pages. 1983.

Vol. 1007: Geometric Dynamics. Proceedings, 1981. Edited by J. Palis Jr. IX, 827 pages. 1983.

Vol. 1008: Algebraic Geometry. Proceedings, 1981. Edited by J. Dolgachev. V, 138 pages. 1983.

Vol. 1009: T. A. Chapman, Controlled Simple Homotopy Theory and Applications. III, 94 pages. 1983.

Vol. 1010: J.-E. Dies, Chaînes de Markov sur les permutations. IX, 226 pages. 1983.

Vol. 1011: J. M. Sigal. Scattering Theory for Many-Body Quantum Mechanical Systems. IV, 132 pages. 1983.

Vol. 1012: S. Kantorovitz, Spectral Theory of Banach Space Operators. V, 179 pages. 1983.

Vol. 1013: Complex Analysis - Fifth Romanian-Finnish Seminar. Part 1. Proceedings, 1981. Edited by C. Andreian Cazacu, N. Boboc, M. Jurchescu and I. Suciu. XX, 393 pages. 1983.

Vol. 1014: Complex Analysis - Fifth Romanian-Finnish Seminar. Part 2. Proceedings, 1981. Edited by C. Andreian Cazacu, N. Boboc, M. Jurchescu and I. Suciu. XX, 334 pages. 1983.

Vol. 1015: Equations différentielles et systèmes de Pfaff dans le champ complexe - II. Seminar. Edited by R. Gérard et J. P. Ramis. V, 411 pages. 1983.

Vol. 1016: Algebraic Geometry. Proceedings, 1982. Edited by M. Raynaud and T. Shioda. VIII, 528 pages. 1983.

Vol. 1017: Equadiff 82. Proceedings, 1982. Edited by H. W. Knobloch and K. Schmitt. XXIII, 666 pages. 1983.

Vol. 1018: Graph Theory, Łagów 1981. Proceedings, 1981. Edited by M. Borowiecki, J. W. Kennedy and M. M. Sysło. X, 289 pages. 1983.

Vol. 1019: Cabal Seminar 79-81. Proceedings, 1979-81. Edited by A. S. Kechris, D. A. Martin and Y. N. Moschovakis. V, 284 pages. 1983.

Vol. 1020: Non Commutative Harmonic Analysis and Lie Groups. Proceedings, 1982. Edited by J. Carmona and M. Vergne. V, 187 pages. 1983.

Vol. 1021: Probability Theory and Mathematical Statistics. Proceedings, 1982. Edited by K. Itô and J.V. Prokhorov. VIII, 747 pages. 1983.

Vol. 1022: G. Gentili, S. Salamon and J.-P. Vigué. Geometry Seminar "Luigi Bianchi", 1982. Edited by E. Vesentini. VI, 177 pages. 1983.

Vol. 1023: S. McAdam, Asymptotic Prime Divisors. IX, 118 pages. 1983.

Vol. 1024: Lie Group Representations I. Proceedings, 1982-1983. Edited by R. Herb, R. Lipsman and J. Rosenberg. IX, 369 pages. 1983.

Vol. 1025: D. Tanré, Homotopie Rationnelle: Modèles de Chen, Quillen, Sullivan. X, 211 pages. 1983.

Vol. 1026: W. Plesken, Group Rings of Finite Groups Over p-adic Integers. V, 151 pages. 1983.

Vol. 1027: M. Hasumi, Hardy Classes on Infinitely Connected Riemann Surfaces. XII, 280 pages. 1983.

Vol. 1028: Séminaire d'Analyse P. Lelong - P. Dolbeault - H. Skoda. Années 1981/1983. Edité par P. Lelong, P. Dolbeault et H. Skoda. VIII, 328 pages. 1983.

Vol. 1029: Séminaire d'Algèbre Paul Dubreil et Marie-Paule Malliavin. Proceedings, 1982. Edité par M.-P. Malliavin. V, 339 pages. 1983.

Vol. 1030: U. Christian, Selberg's Zeta-, L-, and Eisensteinseries. XII, 196 pages. 1983.

Vol. 1031: Dynamics and Processes. Proceedings, 1981. Edited by Ph. Blanchard and L. Streit. IX, 213 pages. 1983.

Vol. 1032: Ordinary Differential Equations and Operators. Proceedings, 1982. Edited by W. N. Everitt and R. T. Lewis. XV, 521 pages. 1983.

Vol. 1033: Measure Theory and its Applications. Proceedings, 1982. Edited by J. M. Belley, J. Dubois and P. Morales. XV, 317 pages. 1983.

Vol. 1034: J. Musielak, Orlicz Spaces and Modular Spaces. V, 222 pages. 1983.

Vol. 1035: The Mathematics and Physics of Disordered Media. Proceedings, 1983. Edited by B. D. Hughes and B. W. Ninham. VII, 432 pages. 1983.

Vol. 1036: Combinatorial Mathematics X. Proceedings, 1982. Edited by L. R. A. Casse. XI, 419 pages. 1983.

Vol. 1037: Non-linear Partial Differential Operators and Quantization Procedures. Proceedings, 1981. Edited by S. I. Andersson and H.-D. Doebner. VII, 334 pages. 1983.

Vol. 1038: F. Borceux, G. Van den Bossche, Algebra in a Localic Topos with Applications to Ring Theory. IX, 240 pages. 1983.

Vol. 1039: Analytic Functions, Błażejewko 1982. Proceedings. Edited by J. Ławrynowicz. X, 494 pages. 1983

Vol. 1040: A. Good, Local Analysis of Selberg's Trace Formula. III, 128 pages. 1983.

Vol. 1041: Lie Group Representations II. Proceedings 1982-1983. Edited by R. Herb, S. Kudla, R. Lipsman and J. Rosenberg. IX, 340 pages. 1984.

Vol. 1042: A. Gut, K. D. Schmidt, Amarts and Set Function Processes. III, 258 pages. 1983.

Vol. 1043: Linear and Complex Analysis Problem Book. Edited by V. P. Havin, S. V. Hruščëv and N. K. Nikol'skii. XVIII, 721 pages. 1984.

Vol. 1044: E. Gekeler, Discretization Methods for Stable Initial Value Problems. VIII, 201 pages. 1984.

Vol. 1045: Differential Geometry. Proceedings, 1982. Edited by A. M. Naveira. VIII, 194 pages. 1984.

Vol. 1046: Algebraic K-Theory, Number Theory, Geometry and Analysis. Proceedings, 1982. Edited by A. Bak. IX, 464 pages. 1984.

Vol. 1047: Fluid Dynamics. Seminar, 1982. Edited by H. Beirão da Veiga. VII, 193 pages. 1984.

Vol. 1048: Kinetic Theories and the Boltzmann Equation. Seminar, 1981. Edited by C. Cercignani. VII, 248 pages. 1984.

Vol. 1049: B. Iochum, Cônes autopolaires et algèbres de Jordan. VI, 247 pages. 1984.

Vol. 1050: A. Prestel, P. Roquette, Formally p-adic Fields. V, 167 pages. 1984.

Vol. 1051: Algebraic Topology, Aarhus 1982. Proceedings. Edited by I. Madsen and B. Oliver. X, 665 pages. 1984.

Vol. 1052: Number Theory. Seminar, 1982. Edited by D.V. Chudnovsky, G.V. Chudnovsky, H. Cohn and M.B. Nathanson. V, 309 pages. 1984.

Vol. 1053: P. Hilton, Nilpotente Gruppen und nilpotente Räume. V, 221 pages. 1984.

Vol. 1054: V. Thomée, Galerkin Finite Element Methods for Parabolic Problems. VII, 237 pages. 1984.

Vol. 1055: Quantum Probability and Applications to the Quantum Theory of Irreversible Processes. Proceedings, 1982. Edited by L. Accardi, A. Frigerio and V. Gorini. VI, 411 pages. 1984.

Vol. 1056: Algebraic Geometry. Bucharest 1982. Proceedings, 1982. Edited by L. Bădescu and D. Popescu. VII, 380 pages. 1984.

Vol. 1057: Bifurcation Theory and Applications. Seminar, 1983. Edited by L. Salvadori. VII, 233 pages. 1984.

Vol. 1058: B. Aulbach, Continuous and Discrete Dynamics near Manifolds of Equilibria. IX, 142 pages. 1984.

Vol. 1059: Séminaire de Probabilités XVIII, 1982/83. Proceedings. Edité par J. Azéma et M. Yor. IV, 518 pages. 1984.

Vol. 1060: Topology. Proceedings, 1982. Edited by L. D. Faddeev and A. A. Mal'cev. VI, 389 pages. 1984.

Vol. 1061: Séminaire de Théorie du Potentiel. Paris, No. 7. Proceedings. Directeurs: M. Brelot, G. Choquet et J. Deny. Rédacteurs: F. Hirsch et G. Mokobodzki. IV, 281 pages. 1984.